AF791526

GENERAL TEXTBOOK OF ENTOMOLOGY

GENERAL TEXTBOOK OF ENTOMOLOGY

Harold Kuzman

Apple Academics

GENERAL TEXTBOOK OF ENTOMOLOGY

Copyright © Apple Academic Press Inc.

All rights reserved. No part of this publication may be reproduced or transmitted in any form or by any means, electronic or mechanical, including photocopying, recording, taping, or any information storage and retrieval system, without permission from the publisher.

Published by
Apple Academic Press Inc.
3333 Mistwell Crescent
Oakville, ON L6L 0A2
Fax: (866) 222-9549
www.appleacademicpress.com

ISBN 978-0-9811660-0-1

Harold Kuzman

First Edition: 2010

9 8 7 6 5 4 3 2 1

Library and Archives Canada Cataloguing in Publication Data
CIP Data on file with the Library and Archives Canada

Printed in India.

CONTENTS

INTRODUCTION

"There is a difference between a grub and a butterfly; yet your butterfly was a grub."

—Shakespeare

In undertaking a study of insects it is well first of all to know something about what they are, their general nature, appearance, habits and development. The insects comprise the largest group of animals on the globe. There are about four times as many different kinds of insects as all other kinds of animals combined. Insects vary greatly in size. Some are as large as small birds, while others are so small that a thousand placed in one pile would not equal the size of a pea.

Insects are commonly spoken of as "bugs." This term, however, is properly used only when referring to the one order of insects which includes the sap and blood-sucking insects such as the chinch bug, bed-bug, squash bug, and the like. Then too, there are many so-called "bugs" which are not insects at all. Spiders, thousand-legs, crawfishes and even earth-worms are often spoken of as bugs.

What They Are

Insects are variously formed, but as a rule the mature ones have three and only three pairs of legs, one pair of feelers, one pair of large eyes, and one or two pairs of wings. The body is divided into a head, thorax and abdomen.

The head bears the eyes, feelers and mouth, the thorax bears the legs and wings, and the abdomen is made up of a number of segments. The presence of wings at once decides whether or not it is an insect, for, aside from bats and birds, insects alone have true wings. These are the distinguishing characters of the full-grown insect, but, like birds, they hatch from eggs and while young do not always look like their parents. When young they may take on various shapes as caterpillars, borers, maggots, grubs, hoppers, and the like. Young insects are often difficult to distinguish from true worms, centipedes, snails, and such forms, but after one has collected and reared some of the young and watched them pass through the different stages and emerge with wings they are much more easily recognized.

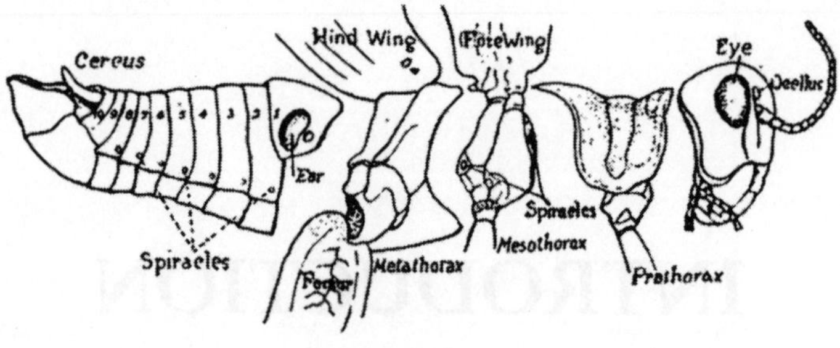

FIG. 1 Side view of grasshopper with wings and legs partly removed. Note the division of the body into head, thorax, composed of prothorax, mesothorax and metathorax, and abdomen consisting of ring-like segments.

Their Principal Characteristics

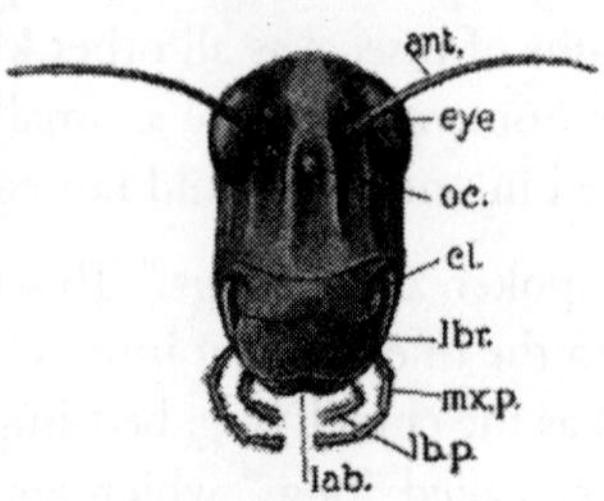

FIG. 2 Face of grasshopper enlarged showing parts; ant., antenna; eye, compound eye; oc., ocellus or simple eye; cl., clypeus; lbr., labrum or upper lip; mx. p., maxillary palpus; lb. p., labial palpus; lab., labium or lower lip.

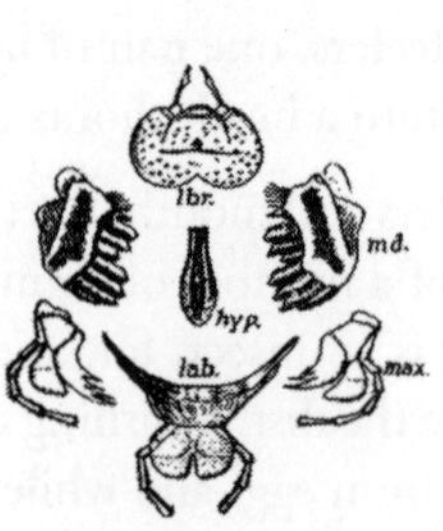

FIG. 3 Mouth parts of grasshopper shown in relative position; lbr., labrum; md., mandibles; hyp., hypopharynx; max., maxillae; lab., labium.

Young insects as a rule are soft like caterpillars and maggots, while the old ones usually have a hard body wall, similar to the beetles and wasps. The wings are usually thin and transparent though in some cases they are leathery or hard as in case of beetles or covered with scales as in the butterflies. The three pairs of legs are jointed and used for running, climbing, jumping, swimming, digging or grasping. The feelers or antennae are usually threadlike, clubbed, or resemble a feather and extend forward or sidewise from the head. The large eyes are compound, being made up of many great small units which, when magnified, resemble honey-comb. In some cases two or three small bead-like eyes may be present besides the two large eyes. The mouth parts of insects may be formed for chewing, as in the grasshopper, or for sucking up liquids, as in the mosquito. The mouth of an insect is built on an entirely different plan from our own. Chewing insects have an upper and lower lip and between these there are two pairs of grinding jaws. These jaws are hinged at the side of the face and when chewing they come together from either side so as to meet in the middle of the mouth. They therefore work sidewise rather than up and down. The mouth parts of the sucking insects are drawn out to form a sucking tube or proboscis as in case of the butterfly or mosquito.

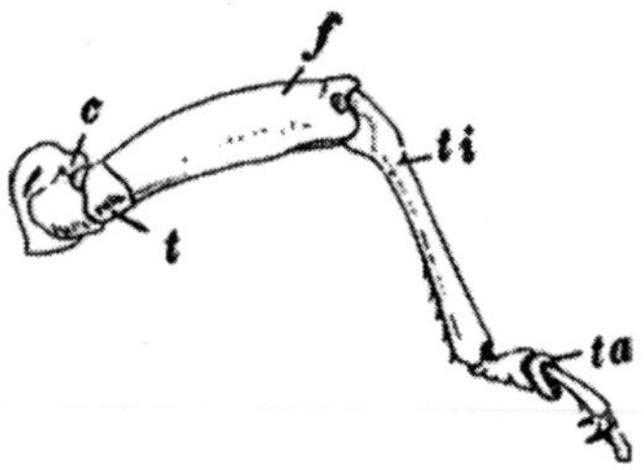

FIG. 4 Leg of grasshopper showing segmentation. The basal segment c, is the coxa, the next t, the trochanter, the large segment f, the femur, the long slender one ti, the tibia, and the three jointed tarsus ta, with claws at the tip.

The internal organs of insects are similar to those of other animals. The digestive tube consists of esophagus, gizzard, or stomach, and intestines. The nervous system is well developed as shown by the extreme sensitiveness of insects to touch. The brain is comparatively small except in the bees and ants. The circulatory system consists simply of a long tube heart, the blood vessels being absent. In this way the internal organs of the insect are simply bathed in the blood. The system of respiration is most complicated. The air is taken in through pores usually along the side of the body and is then carried through fine tracheal tubes to all parts of the body. You cannot drown an insect by putting its head under water, since it does not breathe through its mouth. The muscular system is similar to that of other animals which have the skeleton on the outside.

FIG. 5 The internal organs of the honey bee. Note the strong wing muscles in the thorax. The tube-like heart begins in the head and extends back through the thorax and follows the curve of the abdomen. Below the heart is the digestive tube consisting of the slender esophagus which extends back to the expanded honey stomach, in which the bee carries the nectar it collects from flowers, then the curled true stomach, the small intestine and expanded large intestine. Below this is the nervous system consisting of the brain and a chain of connected enlargements or ganglia extending back into the abdomen in the lower part of the body. The respiratory system in part appears just above the honey stomach, and the black circular or oval spots are cross sections of connecting air tubes, which run all through the body. Also note the sting with the poison gland and sack which are pulled out with the sting; also the sucking tube for getting honey from flowers, and the structures on the legs for gathering and carrying pollen; the pollen basket is on the back side of the hind leg.

Their Methods of Developing

In most cases the parent insect deposits small eggs which hatch later into the young insects. In some cases, as with the blow-flies, the maggot may hatch from the egg while yet in the parent's body, when the active larva is born alive. Whether the egg hatches before or after it is deposited, the young insect continues to develop in one of three ways. It may resemble the parent and simply grow as does a kitten, or it may look somewhat like its parent though smaller and without wings, as the young grasshopper, or it may bear no resemblance whatever to the parent, as the caterpillar which feeds and grows and finally spins a cocoon in which it passes to the resting chrysalis stage and later emerges with wings. The development of insects is therefore extremely complicated.

FIG. 6 The chinch bug showing development with incomplete metamorphosis; a, egg; b, first nymph; c, second nymph; d, third nymph; e, fourth nymph; f, adult winged bug; g, chinch bugs extracting sap from corn plant.

The Principal Orders

In order to study a group of animals which includes so many thousand different kinds it is necessary to divide them into a number of sharply defined divisions or orders. All animal life is naturally grouped into such divisions and subdivisions. Among the insects we at once detect seven large, sharply defined divisions or orders, and ten or more smaller ones. Of these we have first, the two-winged true flies; second, the four-winged butterflies and moths; third, the hard-backed beetles; fourth, the stinging four-winged wasps and bees; fifth, the variously formed sucking insects or true bugs, as chinch bugs and bed-bugs; sixth, the rapid-flying four-winged snake doctors or dragon-flies and, seventh, the hopping forms, the grasshoppers. Besides these we have the various smaller orders of water-loving insects, fleas, etc. The seven groups mentioned above include the majority of our common forms and in the studies to follow we will include only representatives from these orders.

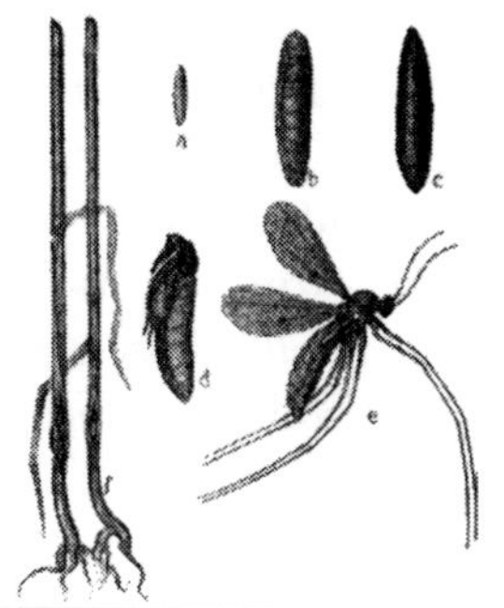

FIG. 7 The Hessian fly showing development with complete metamorphosis; a, egg; b, larva or maggot; c, flax-seed stage; d, pupa; e, adult winged fly; f, wheat stubble with flax-seed stages near base taken after harvest.

Their Habits

The habits of insects are as varied as their forms and adaptations. Some live in the water all their life, others spend a part of their life under water, others live the care-free life of the open air, others enjoy feeding upon and living in the foulest of filth, others associate themselves with certain definite crops or animals thereby doing untold injury, while others produce food and other materials which are to be used by man for his comfort. Every imaginable nook and crook, from the depths of lakes to the tops of mountains, from the warm, sunny south to the cold frigid north, from the foul damp swamps to the heart of our desert lands, offers a home for some small insect.

The most striking habits and developments among insects is found in the more highly advanced families of bees and ants where definite insect societies are formed, resembling in many respects human societies and human activities. Among these villages are established, homes built, battles fought, slaves made, herds kept by shepherds, and even fields cultivated. In these groups we have the nearest approach to human intelligence.

Their Role in Agriculture

Some insects may be very destructive to crops, others are beneficial, while the majority of insects are of no importance to man or agriculture. The various forms of pests such as the chinch bug, potato beetles, and others do an enormous amount of damage each year.

While some insects are extremely injurious, others are very helpful. The products of the honey bee alone amount to millions of dollars a year, to say nothing of its value to the farmer in pollinating fruits. The annual output of silk, all of which is spun by the silkworm, is worth millions of dollars. Many other forms are of value to man by producing material of commercial value such as lac, shellack, dyes, medicines, etc. Of all the beneficial insects those which are of greatest value to man are the parasites and similar forms which help to keep in check many of the severest pests of the farm. Insects are not all bad fellows by any means. One must acquaint himself with those which are good and those which are bad before he can hope to deal with them intelligently.

CHAPTER ONE
SOME COMMON INSECTS

The Grasshopper

The grasshopper or locust is one of the most ancient plagues of cultivated crops. From the earliest time they have destroyed crops. During Moses' sojourn in Egypt they were so destructive as to cause severe famine and various other references to their destructive work are to be found in the early writings. Since those early days, just think of the crops that have disappeared between the greedy jaws of grasshoppers!

FIG. 8 Common differential locust or grasshopper; a, egg nests underground; b, young nymph; c, d, older nymphs; e, adult grasshopper; f, nymph feeding on grass. This shows development with incomplete metamorphosis.

The House Fly or Typhoid Fly

In the house fly we find one of man's most deadly foes. War cannot compare with the campaigns of disease and death waged by this most filthy of all insects.

The fly is dangerous on account of its filthy habits. It breeds in filth, feeds on filth in open closets, slop-barrels, on the streets and in back alleys and then comes into the house and wipes this germ-laden filth on our food or on the hands or even in the mouths of helpless babies. Who has not seen flies feeding on running sores on animals, or on "spit" on sidewalks? These same flies the next minute may be feeding on fruits or other food materials. We rebel when pests destroy our crops or attack our stock, but here we have a pest which endangers our very lives, and the lives of those dear to us.

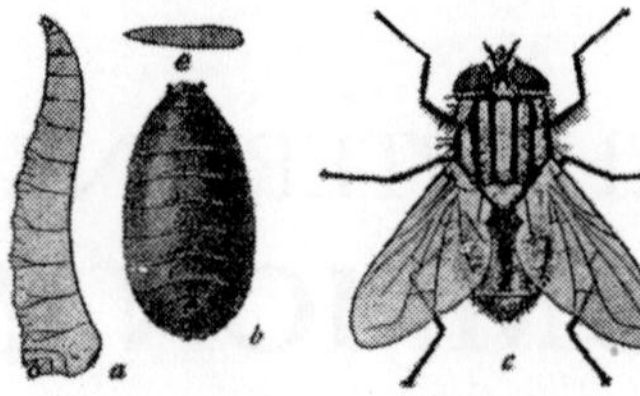

FIG. 9 House fly; a, larva or maggot; b, pupa; c, adult; e, egg. All enlarged

If the fly confined itself to filth we could overlook it as it would help to hasten the removal of filth. On the other hand, if it avoided filth and remained in our home we could not overlook it, but we could feel safe that it was not apt to do us a great deal of harm. But one minute it is here and the next somewhere else; from filth to foods and then back again to filth. In this way it carries disease germs upon its feet and other parts of its body and by coming in contact with food material some of these germs are sure to be left on it and cause trouble later. The fly's method of carrying disease is different from that of the mosquito where the germ is carried inside its body.

The presence of flies in the home is usually a sign of untidiness; but it means more, it means that disease and often death is hovering over the home. We are too apt to consider the fly simply as a nuisance when we should take it more seriously. The child should be led to realize that the fly should not be tolerated in the home, that it is dangerous and that it can and must be destroyed.

The house fly may pass the winter either as the adult fly in cracks and crannies about the home, or in out-buildings or it may remain as a hard, brown, oval pupa in stables and manure piles when, with the first warm days of spring, it escape from this case as the fly ready to lay eggs for the first colony. This ceaseless breeding continues from spring until the first frost in the fall.

In the control of the fly and prevention of trouble from it there are three important steps to take. First of all, go to the source of the trouble and do away with or screen all breeding places. Then, by keeping in mind the fact that the fly is comparatively harmless as long as it is kept from filth laden with germs, do away with all open closets, uncovered slop-barrels and other filth. As a further precaution keep it from the home by the use of screens and when necessary "swatters." Do not make the mistake of trying to control the pest with the "swatters" alone.

In the country too often manure is permitted to accumulate about the barn during the summer with a view of using it on wheat ground in the fall and this furnishes ideal conditions for the fly to breed. Another source of constant danger especially in the rural districts is the presence of open closets or worse still the presence of no closet at all. This is without doubt the most dangerous accessory of the farm. More screens should be used in the home and greater care in keeping them closed.

The Mosquito

Here we have another small insect which, like the house fly, is extremely dangerous, due to its ability to carry the germs of disease. There are hundreds of species of mosquitoes, some small, some large. The majority of these are unable to carry disease so far as we know at present, but they should be avoided as dangerous. The Missouri forms which carry disease are the so-called malarial fever mosquitoes, and they are entirely responsible for the transmission of this sapping and often fatal disease. In the warm countries these are more abundant and the fever is more fatal. In the south there is still another disease-carrying mosquito, the yellow fever mosquito. This form is most dangerous of all.

The mosquito first bites a patient suffering with malaria and in this way it takes in germs along with the blood which it sucks from the patient. After these germs pass through stages of development in the body of the mosquito they are ready to be injected back into a healthy person where, in due time, they cause the disease. The germs feed inside the red blood corpuscles and at regular intervals they destroy a large number of these causing a chill which is followed by fever and a new supply of corpuscles is produced. This alternation of chill and fever may continue all summer, if medicine is not taken to destroy the germs. Quinine will kill the germs if it is taken so that plenty of it is in the blood when the germs come out of the torn down corpuscles during a chill.

In order to prevent malarial fever, get rid of the mosquitoes by draining and oiling the breeding places, escape their bites by screening houses, smudging and destroying the adults, and keep the mosquitoes from patients who have the fever. This is almost as important as the destruction of the mosquitoes. The malarial fever mosquitoes are as harmless as our common forms so long as they do not become infected with germs by sucking blood from a fever patient.

FIG. 10 Mosquitoes in position for biting; a, common Culex mosquito; b, malarial fever mosquito. Note that the one stands parallel, while the other stands at an angle to the surface on which it rests.

In view of the fact that most of our common mosquitoes are classed as non-dangerous, it is of interest to know just how to distinguish the harmless ones from the dangerous. The adults of the two forms can be easily distinguished when they are seen at rest. The common forms always rest with the body parallel to the surface on which they rest, while the malarial form always elevates the end of the body so that the head is pointed toward the surface on which it rests. In like manner the wigglers can be distinguished from each other. Our common wigglers always hang head downward in the water while those of the malarial mosquitoes rest near the surface of the water with their bodies parallel to it. The majority of the wigglers found in rain barrels are of our common forms.

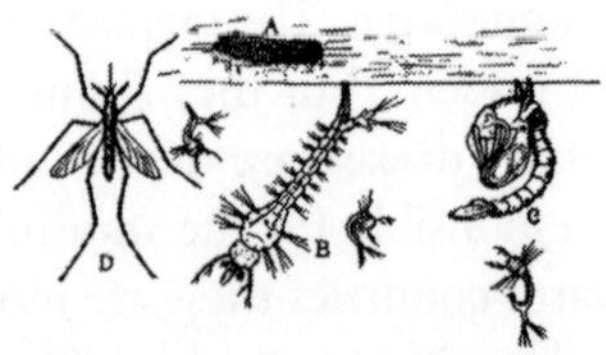

FIG. 11 Common Culex mosquito showing stages of development; a, raft of eggs; b, larvæ or wigglers of different sizes; c, pupa; d, mosquito. The large wiggler and the pupa are taking air from the surface of the water through their breathing tubes.

The life of the mosquito is quite interesting and is an excellent example of an insect which lives in the water part of its life and in the air the rest. The mature female mosquito, which does all the biting, searches for water in rain barrels, cans, ditches, ponds, and stagnant swamps where she lays her eggs either in raft-shaped packets or singly. When the wigglers hatch they swim about in the water and feed upon decaying material and microscopic water plants. When the wiggler is full grown it changes to an active pupa which has a large head and a slender tail and is more or less coiled. A little later the winged mosquito escapes. In the rural districts most of the mosquitoes breed in stagnant ponds, swamps and rain barrels and from these they fly to the home where they cause trouble. Such places should be drained or protected with oil or other means to prevent the mosquito from using them for breeding purposes. Ponds can be freed of the wigglers by introducing fish or by using a small amount of coal oil on the surface. The wigglers have a breathing tube which is thrust out above the water when fresh air is needed and if there is a thin film of oil on the water this is prevented. Rain barrels can be freed of the pest in this way also, or perhaps better by covering them with a cloth. The mosquitoes are most troublesome about the home at night. When one sits out doors he should keep a smudge going to drive them away while screens will keep them out of the house at night.

The Cabbage Miller

With the first approach of spring comes swarms of large green flies which bask in the March sun on the south sides of buildings. They are not with us long, however, until we notice flashes of white quickly moving about from one early weed to another. These are the advance guards of the cabbage millers or butterflies. All through the cold winter they remained in the chrysalis stage stuck to the sides of houses, fence posts and in other protected places, awaiting the first breath of spring. The first adults to emerge find no cabbage on which to lay their eggs so they are compelled to use other plants such as pepper grass.

The eggs are very small and are usually placed on the lower edge of the leaf. These hatch and the small green worms appear. Throughout the summer there are a number of broods produced and an enormous amount of damage is done. Just before frost the last caterpillars search for protected places where they pass to the pupal or resting stage for the winter. No cocoon is spun by this caterpillar.

Where measures are not taken to control the cabbage worms they destroy much of the cabbage crop each season. The white butterflies can be seen any day during the summer visiting cabbage, mustard, radishes and other similar plants. By destroying all of the worms and millers in the early spring one has less trouble later. This can be done by hand picking, or where the patch is large by spraying with a poison solution to which soap is added to keep the solution from rolling off in large drops. Poison can be used until the heads are well formed, but if the first worms in the spring are destroyed, later spraying is unnecessary though an occasional handpicking will help.

FIG. 12 Cabbage worm feeding, slightly enlarged.

FIG. 13 Pupa or chrysalis of cabbage miller.

FIG. 14 Cabbage miller on red clover blossom.

The Apple Worm

This is perhaps the most destructive insect pest attacking the apple. It is most destructive to the apple, though the pear comes in for its share. The small worms crawl down in the blossom end of the young developing apple and from there bore into the pulp and eventually reach the core of the fruit. They stay in the apple about six weeks when they eat a hole out to the surface and crawl down to the trunk where loose bark offers a hiding place. Here they spin their cocoons and change to a small, brown, plump pupa and after a few days the winged moth emerges. The moth is very small and is not often found by one not acquainted with it. They come out as the temperatures grow warmer, when they lay eggs for a second colony of worms which again enter the fruit and destroy more of it. These worms of the second brood are usually mature and leave the fruit about the time apples are picked. They escape and soon spin cocoons in which they pass the winter. Early in the spring these change to pupae and later the moths come out. They appear about the time apples bloom in the spring and lay the eggs for the first worms which enter in great numbers at the blossom end.

FIG. 16 The codling moth slightly enlarged.

FIG. 15 *Apple worms in core of apple. Usually only one worm appears in an apple.* Note the decaying of the apple.

The Tomato or Tobacco Worm

This insect is often very destructive to tomatoes and tobacco. When full grown it is four inches long, usually dark green with a number of slanting white lines along either side. It is so near the color of the plants that it is difficult to see it.

During the summer months the worms are common, being most abundant in August. In the fall the mature worms go into the ground and change from the worm to a large, oval, brown pupa with a jug-handle-like appendage on the under side. These are often turned up when the garden is plowed in the spring. After tomato plants are well started the large greyish humming-bird-like moths comes from the ground and begin laying eggs. The moth expands from four to six inches and is often seen at dusk visiting the blossoms of "jimson weed" and other large tube flowers. They are also found around lights at night.

FIG. 17 Young tomato worm.

FIG. 18 Tomato moth.

Where they are troublesome the plants should either be sprayed with a poison when the injury is first noticed or else the worms should be picked off and destroyed. There is a small parasitic wasp which is very helpful in destroying this caterpillar. They live inside the worm and when mature bore out through the skin on the sides and back where they spin small white egg-like cocoons from which later the small wasps emerge. Often a hundred or more may come out of one worm.

The Firefly

This insect is of little economic importance to us at present but its peculiar habit of producing light makes it a very striking form and one which deserves study. The firefly is a beetle. They live largely in damp places and bottoms at night are specked with their tiny flashes of light. The larval or grub stage is passed on the ground beneath grass, weeds and rubbish where they often prey upon other insects. In some cases the grubs may be able to produce light though as a rule the luminous grub-like creature or glow-worm is a wingless adult firefly.

FIG. 19 Firefly beetles on sour-dock leaf.

FIG. 20 Lower surface of firefly beetle enlarged to show the light producing segments of the abdomen.

The White Grub or June-Bug

This insect is more familiar to country children in the grub stage. Every one who has followed a plow in rich sod land has seen these fat, white coiled grubs roll down into the furrow when the plow turns them up. They are in the ground feeding on the roots of plants. Often all the roots of grass in lawns and meadows are eaten off and the sod dies and can be rolled up like strips of carpet. This insect breeds largely in sod and when this is plowed under and other crops are planted the grubs may injure them severely.

FIG. 21 White grub feeding on roots of corn plant, enlarged.

Corn, wheat, oats and truck crops are severely injured. In some cases the grubs may feed for three years before they change to the pupa and later to the adult beetle. To control this pest, plow in the fall and rotate crops, so that sod will not remain on the same land too long.

FIG. 22 June beetle showing stages of development. The larva or grub worm is feeding on roots under ground; the pupa and adult are shown above ground.

The beetles come from the ground or may be plowed out in May and June and are commonly spoken of as May-beetles or June-bugs. They are usually of a yellowish-brown color and are often troublesome coming into the house at night where they buzz about the light, bumping into everything until they finally drop heavily to the floor.

The Potato Beetle

This is one of the worst pests of the potato. The full-grown insect is short and thick with a hard shell, striped with yellow and black. The grubs, on the other hand, are soft and red or orange with black spots. Both the grubs and the beetles feed on potatoes and often completely strip them of their leaves. Since they feed on foliage they can be very quickly destroyed by dusting or spraying the plants with a poison such as Paris green or arsenate of lead. The patches of yellow or reddish eggs are found mostly on the under side of the potato leaves. When the fat grubs are full-fed they go into the ground and change to pupae and later to the striped beetles. This pest should not be mistaken for the so-called old-fashioned potato beetles which are long and slender and either bluish grey in color or striped with yellow and black. These are blister beetles and are entirely different.

FIG. 23 The potato beetle showing stages of development and work on a potato plant. Note the small patch of eggs and different sized grub on the plant and the grub, pupa and adult at side.

The Lady-Beetle (or Lady Bug)

"Hurt no living thing:
Ladybird, nor butterfly,
Nor moth with dusty wing,
Nor cricket chirping cheerily,
Nor grasshopper so light of leap,
Nor dancing gnat, nor beetle fat,
Nor harmless worms that creep."
—Christina Rossetti.

The lady-beetles comprise one family of small beetles, which is famous for the number of beneficial forms it includes. Most forms feed upon other insects, in most cases pests such as plant-lice and scale insects. From the time they hatch from the egg until they pupate and again after the beetle stage is reached they are regular tigers after plant-lice. They catch and hold their prey between the front feet while they devour it bodily. The larva of the lady-beetle has an astonishing capacity for in one day it will eat several times its own weight of plant-lice. Farmers and fruit growers could hardly get along without the help of these small beetles and yet unfortunately thousands are often destroyed by those who do not know of their beneficial work.

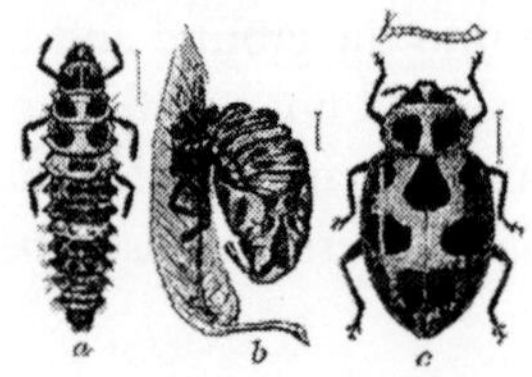

FIG. 24 The spotted lady-beetle; a, larva; b, pupa; c, adult; enlarged.

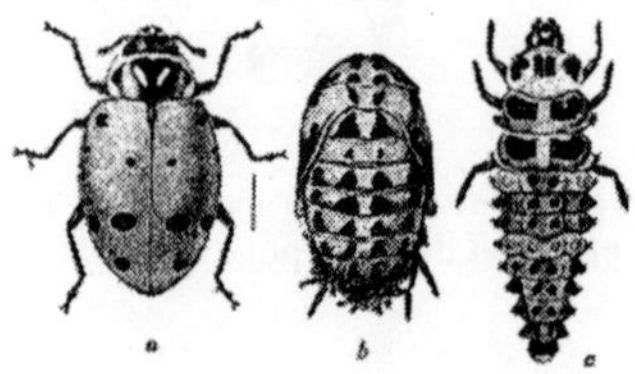

FIG. 25 The convergent lady-beetle; a, adult; b, pupa; c, larva; all enlarged.

The lady-beetles, or lady-bugs as they are often called, are fairly uniform in shape and color. They are oval or round in outline with the back rounded or elevated and the underside flat. In color they are usually either orange or yellow, checkered or blotched with black or black with yellow or bright orange markings. They closely resemble small tortoises. Unfortunately several plant feeding beetles are similar in shape and color which casts reflections on the lady-beetles.

The grub of the lady-beetle is usually black or dull colored with red or yellow markings which make it very conspicuous. It runs about over foliage and is broad in front and tapers to a point behind. When the grub is full fed it attaches the top of its body to a leaf, twig or other object and pupates. In the pupal stage it is often protected with spines and is able to lift the front end of the body up and down when disturbed, producing a light tapping sound.

The lady-beetle usually hides in rubbish about the base of trees or in some cases even enter homes for the winter months, coming out with the spring to deposit small masses of oval yellow or orange eggs on plants infested with lice. They breed rapidly and with the help of parasites and other beneficial insects usually control the plant-lice pests.

The Dragon-Fly

Every lover of the stream or pond has seen these miniature airplanes darting now here, now there but ever retracing their airy flight along the water's edge or dipping in a sudden nose dive to skim its very surface. At times it is seen to rest lazily, wings out-stretched, perched on some projecting reed or other object. But when approached how suddenly it "takes off" and is out of reach. The dragon-fly is an almost perfect model of the modern monoplane. Its two long wings on either side are the planes, its head the nose, its thorax the fuselage and its long projecting abdomen the tail or rudder. On wing the dragon-fly is one of the swiftest and most powerful insects. The dragon-flies are found all over the world being most abundant in the warmer regions where rainfall and bodies of water are abundant. For breeding they require water, their immature stages living under water feeding on aquatic animal life. Our present order of dragon-flies is the remains of an ancient race of insects of immense size. From fossil remains we learn that ancient dragon-flies had a wing expanse of three feet.

FIG. 26 Cast off skin of dragon-fly nymph, showing shape and position taken on a twig when the adult winged form emerged from the last nymph stage.

The dragon-fly is a beneficial insect throughout life. The young feed on mosquito wigglers and similar life in ponds and streams while the adults dart here and there over ponds, fields or lawn catching mosquitoes and other winged insects. Many look upon the dragon-fly as a dangerous stinging insect but it is entirely harmless and can be handled without the least danger. They vary greatly in size and appearance. The so-called damsel-flies form a group of dragon-flies or Odonata which rest with the wings in a vertical position and the young aquatic stages are more slender. In color markings dragon-flies include all hues of the rainbow tho as a rule they do not have such extravagant colors as the butterflies.

FIG. 27 One of our common dragon-flies found about ponds and streams.

The Squash Bug

This common blackish or earth-colored bug is usually called the squash stink-bug. It has a very disagreeable odour which gives it this name. When disturbed it throws off from scent glands a small quantity of an oily substance which produces this odour. This is a protection to it for few birds or animals care to feed on it. Most species of sap or blood sucking true bugs have a similar protecting odor.

FIG. 28 Pumpkin in field covered with adults and nymphs of squash stink-bug.

The squash bug feeds largely on squash and pumpkins. It has a slender beak with needle-like mouth parts which are stuck into the plant for extracting the sap. It feeds only on plant sap. When it can not get squash or pumpkins it will feed on watermelons, muskmelons and related crops. It is very destructive to these crops. It not only extracts sap thus weakening the plant but it also seems to poison the plant while feeding. In this way its bite injures the plant something like the effects of the bed-bug's bite on our flesh. It feeds first on the leaves and vines often killing them in a few days. Later it may cluster and feed on the unripe squashes or pumpkins in such numbers as to completely cover them.

The squash bug lives thru the winter as the matured winged insect. It flies from its food plant to winter quarters late in the fall. For winter protection it may enter buildings, hide under shingles on roofs, crawl into piles of lumber, under bark of dead trees or stumps or hide under any similar protection. When its chosen food crops begin to come up in the spring it leaves its winter home and flies in search of food. After feeding for a time the female lays patches of oval, flattened, gold-colored eggs set on edge. When first deposited the eggs have a pale color but in a short time the golden color appears. In some cases only three or four eggs may be found in one patch while again there may be twenty or thirty of them. They are so brightly colored that they can easily be seen and most boys and girls have seen them on the leaves of squashes or pumpkins.

In a few days after they are laid they hatch and out of each crawls a small, long-legged blackish or greenish young bug called the nymph. These little fellows usually stay in a crowd hiding on the under side of a leaf. After feeding for a time their leaf begins to turn yellow and soon dies. Then they move to a new leaf. As they feed they grow rapidly and after shedding their skins they change to the second nymph stage. This shedding of their skins or molting occurs five times before they mature. Of course each time before the old skin or suit of clothes is discarded a new one is developed beneath. The females may continue to deposit eggs for later clusters of young. They become most abundant on the crop late in the fall. Just before cold weather sets in the adults again seek winter shelter.

This is a very difficult insect to control. Since it feeds on liquid sap only it is impossible to kill it by spraying the crop with a poison such as arsenate of lead. It cannot chew and swallow such poison. The young can be killed fairly well with a spray or dust containing nicotine but such treatments are not effective against the adults or nearly mature nymphs. A better method is to destroy all the bugs possible in the fall before they go to the winter protection and then watch for and destroy the adults and the eggs masses in the spring when they appear on the young crop. If the first adults and the eggs and newly hatched nymphs are destroyed the crop can be protected against the destructive work later.

FIG. 29 Squash stink-bug adult and nymph extracting sap from squash.

The Plant-Louse

The plant-louse or aphis is a sap-sucking insect which feeds and multiplies rapidly often seriously injuring crops. The loss of sap together with the poisoning effect of the bite causes the weakening of the plant or leaf with its ultimate death if feeding continues. The greatest damage is usually done during cold springs or during a cool rainy period. This prevents the enemies of the louse from increasing and attacking it while the weather may not be too severe to prevent the louse from working. Under favorable climatic conditions the natural enemies of the louse as a rule are able to hold it in check. The principal enemies of the louse are certain small insect feeding birds, lady-beetles, syrphid-flies, lace-wings and tiny wasp parasites. The beneficial work of the lady-beetles is discussed in an earlier chapter. The birds and lady-beetles devour them bodily, the larvæ of the lace-wings and syrphid-flies extract their blood while the wasps live as internal parasites.

FIG. 30 Black winter eggs of Aphis showing how they are deposited in masses on twigs of a tree.

In temperate latitudes, the plant-lice as a rule live thru the winter in the form of a fertile egg attached to the twigs of trees and shrubs. The winter egg is produced by a true female plant-louse. As a rule there is only one generation of true males and females produced each year. This brood develops late in the fall to produce the fertilized winter eggs. In the spring these eggs hatch and the tiny nymphs begin to extract sap. On maturing they begin to give birth to young lice. Throughout the summer this method of reproduction continues. These summer forms are known as the stem mothers or agamic females. These are not true females for they produce living young in place of eggs and during the summer no male lice are produced at all. This is nature's way of increasing the race of plant-lice rapidly. Late in the fall again a brood of true males and females is produced. During the summer the plant-lice increase more rapidly than any other type of insect.

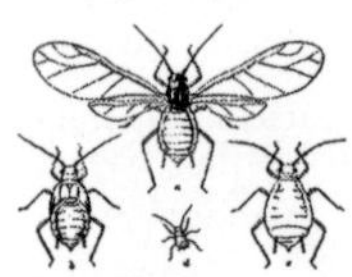

FIG. 31 Plant-lice vary in size, color and general appearance. Many are green while some are red or black.

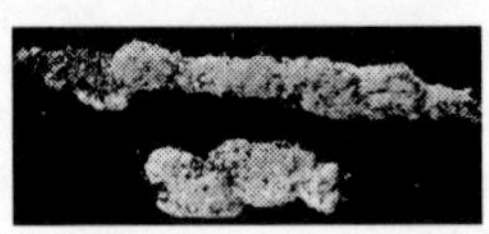

FIG. 32 The cottony secretion produced by plant-lice.

The Honey Bee

"Simple and sweet is their food; they eat no flesh of the living."
—Von Kuebel.

One can hardly believe that this small, ever busy creature each year gathers many million dollars worth of products for man in this country alone to say nothing of its inestimable value on the farm and especially in the orchard, where it assists in carrying pollen from blossom to blossom. It is of far greater value to man as a carrier of pollen than it is as a honey gatherer and yet under especially favorable conditions in one year a strong colony may produce between twenty-five and thirty dollars worth of honey.

FIG. 33 Worker, queen and drone honey bees; all about natural size.

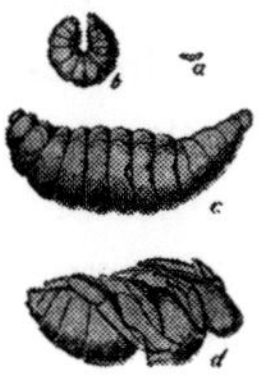

FIG. 34 Stages of development of honey bee; a, egg; b, young grub; c, full-fed grub; d, pupa; all enlarged.

The general habits of the bee are fairly well known by all. They live in colonies consisting largely of workers, one female or queen and males or drones. Whenever the number of workers becomes sufficiently large to warrant a division of the colony, a young queen is reared by the workers and just before she matures, the old queen leaves with about half of the workers to establish a new colony. This division of the colony is called swarming. If a hive, box or other acceptable home is not provided soon after the swarm comes out and clusters, it may fly to the woods and establishes itself in a hollow tree where the regular work of honey gathering is continued. This accounts for so many bee-trees in the woods. The bee has been handled by man for ages, but it readily becomes wild when allowed to escape to the woods.

The bee colony offers one of the best examples to show what can be accomplished by united effort where harmony prevails. Certain of the workers gather honey, others are nurses for the queen and young brood in the hive, others guard the hive and repel intruders, and others care for the hive by mending breaks and providing new comb as it is needed. Each knows its work and goes about it without interfering with the work of others. It is one huge assemblage of individuals under one roof where harmony and industry prevail.

Throughout the long, hot summer days the workers are busy from daylight until dark gathering nectar, while at night they force currents of air thru the hive to evaporate the excess water from the nectar. When flowers are not available near the hive they simply fly until they find them, be it one, two or more miles. As long as they are able to gather honey they continue to do so and when they give out they drop in the field and are forgotten, others rushing to take their place. Often when winter is approaching and the store of honey is low the less vigorous ones are cast out from the hive and left to die. If man could learn a few of the lessons which the bee teaches, he would be a better, a more useful and a wiser addition to society.

The Ant

Ants are closely related to the bees and are similar to them in many respects. They live in colonies consisting of workers, drones, and a queen. The males or drones appear at swarming time and the workers are divided into various castes—warriors, guards, nurses, etc. Those families of ants, however, which seem to have what approaches real intelligence, far outstrip the bees in many respects. In some cases ants seem to be able to plan and carry out lines of work very much the same as man does. The various stages of human intelligence or races of men from the savage to the intelligent man are in a way similar to the various races of ants. There are ants which live as hunters, others which live as shepherds and still others more highly developed which grow crops either in or near the nest as is the case with the fungus growing ants. This striking similarity between the development of ants and man offers ground for much speculation.

FIG. 35 Ant hill showing activity and stages of development; a, egg; b, young grub; c, pupa; d, worker; e, queen with wings; f, worker carrying young grub; all enlarged. The ant hill and workers at work much reduced.

Some ants may be of considerable value to man while others are the source of great annoyance and injury. The tidy housewife usually places the ant in the same category with cockroaches and bed-bugs and the corn growers attribute much of the injury to young corn to the work of the small cornfield ant which acts as a shepherd of the corn root-louse. Ants are usually more destructive by protecting and caring for other pests than by attacking the crop direct.

Every country person is familiar with ants. Ants are as numerous as the stars in the skies and vary in size. They are found from the tropics to the frozen north, in deserts, swamps and in fact, almost any place where plants or animals live. They do not waste time building or manufacturing a complicated nest like wasps and bees, so when food is scare, or for other reasons they need to move they simply "pack up" and migrate. This, together with the fact that they feed on almost every imaginable kind of plant and animal material, accounts in part for the fact that they are the rulers of the insect world.

CHAPTER TWO
GROWTH AND CHANGE

The caterpillar differs markedly from the butterfly. As we pursue our studies of insect growth and transformation we shall find that in some cases the difference between young and adult is much greater—as for example between the maggot and the house-fly, in others far less—as between the young and full-grown grasshopper or plant-bug. It is evidently wise to begin a general survey of the subject with some of those simpler cases in which the differences between the young and adult insect are comparatively slight. We shall then be in a position to understand better the meaning of the more puzzling and complex cases in which the differences between the stages are profound.

In the first place it is necessary to realise that the changes which any insect passes through during its life-story are essentially accompaniments of its growth. The limits of this little book allow only slight reference to features of internal structure; we must be content, in the main, to deal with the outward form. But there is an important relation between this outward form and the underlying living tissues which must be clearly understood. Throughout the great race of animals—the Arthropoda—of which insects form a class, the body is covered outwardly by a *cuticle* or secretion of the underlying layer of living cells which form the outer skin or *epidermis.* This cuticle has regions which are hard and firm, forming an *exoskeleton*, and, between these, areas which are relatively soft and flexible. The firm regions are commonly segmental in their arrangement, and the intervening flexible connections render possible accurate motions of the exoskeletal parts in relation to each other, the motions being due to the contraction of muscles which are attached within the exoskeleton.

Now this jointed exoskeleton—an admirably formed suit of armour though it often is—has one drawback: it is not part of the insect's living tissues. It is a cuticle formed by the solidifying of a fluid secreted by the epidermal cells, therefore without life, without the power of growth, and with only a limited capacity for stretching. It follows, therefore, that at least during the period through which the insect continues to grow, the cuticle must be periodically shed. Thus in the life-story of an insect or other arthropod, such as a lobster, a spider, or a centipede, there must be a succession of cuticle-castings—'moults' or *ecdyses* as they are often called.

When such a moult is about to take place the cuticle separates from the underlying epidermis, and a fluid collects beneath. A delicate new cuticle is then formed in contact with the epidermis, and the old cuticle opens, usually with a slit lengthwise along the back, to allow the insect in its new coat to emerge. At first this new coat is thin and flabby, but after a period of exposure to the air it hardens and darkens, becoming a worthy and larger successor to that which has been cast. The cuticle moreover is by no means wholly external. The greater part of the digestive canal and the whole air-tube system are formed by inpushings of the outer skin (ectoderm) and are consequently lined with an extension of the chitinous cuticle which is shed and renewed at every moult.

In all insects these successive moults tend to be associated with change of form, sometimes slight, sometimes very great. The new cuticle is rarely an exact reproduction of the old one, it exhibits some new features, which are often indications of the insect's approach towards maturity. Even in some of those interesting and primitive insects the Bristle-tails (Thysanura) and Spring-tails (Collembola), in which wings are never developed, perceptible differences in the form and arrangement of the abdominal limbs can be traced through the successive stages. But the changes undergone by such insects are comparatively so slight, that the creatures are often known as 'Ametabola' or insects without transformation in the life-history. Now there are a considerable number of winged insects—cockroaches and grasshoppers for example—in which the observable changes are also comparatively slight. We will sketch briefly the main features of the life-story of such an insect.

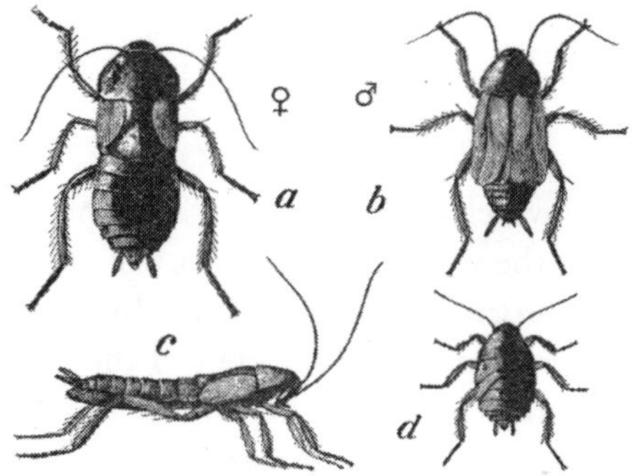

FIG. 36 Common Cockroach (Blatta orientalis). a, female; b, male; c, side view of female; d, young.

The young creature is hatched from the egg in a form closely resembling, on the whole, that of its parent, so that the term 'miniature adult' sometimes applied to it, is not inappropriate. The baby cockroach is known by its flattened body, rounded prothorax, and stiff, jointed tail-feelers or cercopods; the baby grasshopper by its strong, elongate hind-legs, adapted, like those of the adult, for vigorous leaping. During the growth of the insect to the adult state there may be four or five moults, each preceded and succeeded by a characteristic instar. The first instar differs, however, from the adult in one conspicuous and noteworthy feature, it possesses no trace of wings. But after the first or the second moult, definite wing-rudiments are visible in the form of outgrowths on the corners of the second and third thoracic segments. In each succeeding instar these rudiments become more prominent, and in the fourth or the fifth stage, they show a branching arrangement of air-tubes, prefiguring the nervures of the adult's wing. (After the last moult the wings are exposed, articulated to the segments that bear them, and capable of motion. Having been formed beneath the cuticle of the wing-rudiments of the penultimate instar, the wings are necessarily abbreviated and crumpled. But during the process of hardening of the cuticle, they rapidly increase in size, blood and air being forced through the nervures, so that the wings attaining their full expanse and firmness, become suited for the function of flight.

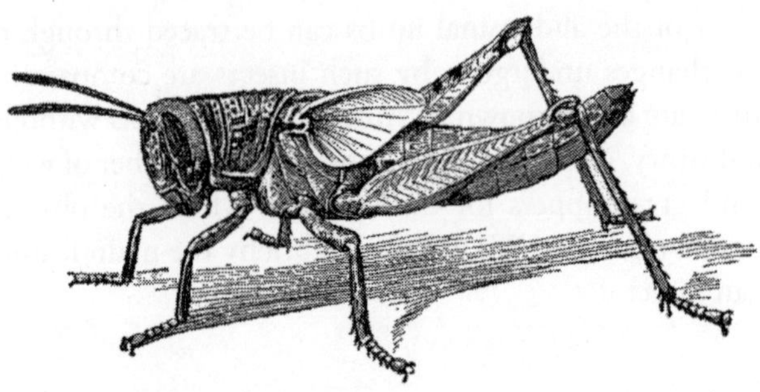

Fig. 37 Nymph of Locust (Schistocera americana) with distinct wing-rudiments.

The changes through which these insects pass are therefore largely connected with the development of the wings. It is noteworthy that in an immature cockroach the entire dorsal cuticle is hard and firm. In the adult, however, while the cuticle of the prothorax remains firm, that of the two hinder thoracic and of all the abdominal segments is somewhat thin and delicate on the dorsal aspect. It needs not now to be resistant, because it is covered by the two firm forewings, which shield and protect it, except when the insect is flying. There are, indeed, slight changes in other structures not directly connected with the wings.

In a young grasshopper, for example, the feelers are relatively stouter than in the adult, and the prothorax does not show the specifically distinctive shape with its definite keels and furrows. Changes in the secondary sexual characters may also be noticed. For instance, in an immature cockroach both male and female carry a pair of jointed tail-feelers or cercopods on the tenth abdominal segment, and a pair of unjointed limbs or stylets on the ninth. In the adult stage, both sexes possess cercopods, but the males only have stylets, those of the female disappearing at the final moult.

Reviewing the main features of the life-story of a grasshopper or cockroach, we notice that there is no marked or sudden change of form. The newly-hatched insect resembles generally its parent, except that it has no wings. Wing-rudiments appear, however, in an early instar as visible outgrowths on the thoracic segments, and become larger after each moult. All through its various stages the immature insect—*nymph* as it is called—lives in the same kind of situations and on the same kind of food as its parent, and it is all along active and lively, undergoing no resting period like the pupal stage in the transformation of the butterfly.

One interesting and suggestive fact remains to be mentioned. There are grasshoppers and cockroaches in which the changes are even less than those just sketched, because the wings remain, even in the adult, in a rudimentary state (as for example in the female of the common kitchen cockroach, *Blatta orientalis*), or are never developed at all. Such exceptional winglessness in members of a winged family can only be explained by the recognition of a life-story, not merely in the individual but in the race. We cannot doubt that the ancestors of these wingless insects possessed wings, which in the course of time have been lost by the whole species or by the members of the female sex. It is generally assumed that this loss has been gradual, and so in many cases it probably may have been. But there are species of insects in which some generations are winged and others wingless; a winged mother gives birth to wingless offspring, and a wingless parent to young with well-developed wings. Such discontinuity in the life-story of a single generation forces us to recognise the possibility of similar sudden mutations in the course of that age-long process of evolution to which the facts of insect growth, and indeed of all animal development, bear striking testimony.

CHAPTER THREE
THE LIFE-STORIES OF SOME SUCKING INSECTS

We may now turn our attention to some examples of the remarkable alternation of winged and wingless generations in the yearly life-cycle of the same species, mentioned at the end of the last chapter. Cockroaches and grasshoppers belong to an order of insects, the Orthoptera, characterised by firm forewings and biting jaws; in all of them the change of form during the life-history is comparatively slight. A great contrast to those insects in the structure of the mouth-parts is presented by the Hemiptera, an order including the bugs, pond-skaters, cicads, plant-lice, and scale-insects. These all have an elongated, grooved labium projecting from the head in form of a beak, within which work, to and fro, the slender needle-like mandibles and maxillae by means of which the insect pierces holes through the skin of a leaf or an animal, and is thus enabled to suck a meal of sap or blood, according to its mode of life. In many Hemiptera—the various families of bugs both aquatic and terrestrial, for example—the life-history is nearly as simple as that of a cockroach. It is the family of the plant-lice (Aphidae) that affords typical illustrations of that alternation of generations to which reference has been made.

In late autumn tiny wingless males and females are found in large numbers on the withered leaves. The sexes pair together, and the females lay their relatively large, smooth, hard-coated black eggs on the twigs; these resistant eggs carry the species safely over the winter. At springtide, when the leaves begin to sprout from the opening buds the aphid eggs are hatched, and the young insects after a series of moults, through which hardly any change of form is apparent, all grow into wingless 'stem-mothers' much larger than the egg-laying females of the autumn. The stem-mothers have the power, unusual among animals as a whole, but not very infrequent in the insects and their allies, of reproducing their kind without having paired with a male. Eggs capable of parthenogenetic development, produced in large numbers in the ovaries of these females, give rise to young which, developing within the body of the mother, are born in an active state. Successive broods of these wingless virgin females appear through the spring and summer months, and as the rate of their development is rapid, often the whole life-story is completed within a week. The aphid population increases very fast. Later a generation appears in which the thoracic segments of the nymphs are seen to bear wing-rudiments like those of the young cockroach, and a host of winged females are produced; these have the power of migrating to other plants. We understand that wings are not necessary to the earlier broods whose members have plenty of room and food on their native shoots, but that when the population becomes crowded, a winged brood capable of emigration is advantageous to the race.

Many generations of virgin female aphids, some wingless, others winged when adult, succeed each other through the summer months. At the close of the year the latest brood of these bring forth young, which develop into males and egg-laying females; thus the yearly cycle is completed. Variations in points of detail may be noticed in different species of aphids. The autumn males and egg-laying females are, for example, frequently winged, and the same species may have constantly recurring generations of different forms adapted for different food-plants, or for different regions of the same food-plant. But taking a general view of the life-story of aphids for comparison with the life-story of other insects, three points are especially noteworthy. Virgin reproduction recurs regularly, parthenogenetic broods being succeeded by a single sexual brood. A winged parent brings forth young which remain always wingless, and wingless adults produce young which acquire wings. The wings are developed, as in the cockroach, from outward and visible wing-rudiments.

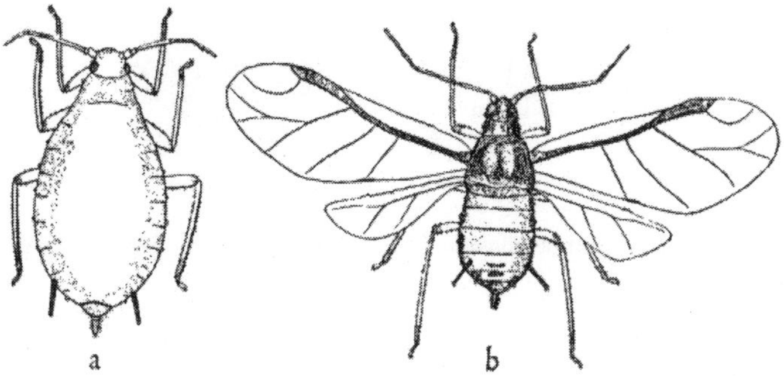

FIG. 38 Apple Aphid (Aphis pomi), virgin females, a, wingless; b, winged.

A family of Hemiptera, related to the Aphidae and equally obnoxious to the gardener, is that of the Coccidae or scale-insects. These furnish an excellent illustration of features noticeable in certain insect life-histories. In the first place, the newly-hatched young differs markedly from the parent in the details of its structure. A young coccid is flattened oval in shape, has well-developed feelers and legs, and runs actively about, usually on the leaves or bark of trees and shrubs, through which it pierces with its long jaws, so that it may suck sap from the soft tissues beneath. After a time it fixes itself by means of these jaws and the characteristic scale or protective covering, composed partly of a waxy secretion and partly of dried excrement, begins to grow over its body. The female loses legs and feelers, and never acquires wings, becoming little more than a sluggish egg-bag. The male on the other hand passes into a second larval stage in which there are no functional legs, but rudiments of legs and of wings are present on the epidermis beneath the cuticle. The penultimate instar of this sex in which the wing-rudiments are visible externally lies passively beneath the scale, its behaviour resembling that of a butterfly pupa. The adult winged male leads a short, but active life.

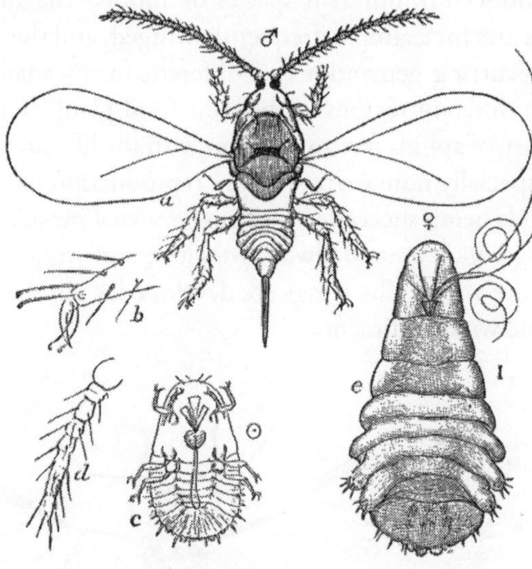

FIG. 39 Mussel Scale-insect (Mytilaspis pomorum). a, male; b, foot of male; c, larva, ventral view; d, feeler of larva; e, female, ventral view.

Another family allied to the Aphidae is that of the Cicads, hardly represented in our fauna but abundant in many of the warmer regions of the earth. Here also the young insect differs widely from its parent in form, living underground and being provided with strong fore-legs for digging in the soil. After a long subterranean existence, usually extending over several years, the insect attains the penultimate stage of its life-story, during which it rests passively within an earthen cell, awaiting the final moult, which will usher in its winged and perfect state.

In the life-histories of cicads and coccids, then, there are some features which recall those of the caterpillar's transformation into the butterfly. The newly-hatched insect is externally so unlike its parent that it may be styled a larva. The penultimate instar is quiescent and does not feed. But while the caterpillar shows throughout its life no outward trace of wings, external wing-rudiments are evident in the young stages of the cicad. In the male coccid we find a late larval stage with hidden wing-rudiments, the importance of which, for comparison with the caterpillar, will be appreciated later.

CHAPTER FOUR
FROM WATER TO AIR

Insects as a whole are preeminently creatures of the land and the air. This is shown not only by the possession of wings by a vast majority of the class, but by the mode of breathing to which reference has already been made, a system of branching air-tubes carrying atmospheric air with its combustion-supporting oxygen to all the insect's tissues. The air gains access to these tubes through a number of paired air-holes or spiracles, arranged segmentally in series.

It is of great interest to find that, nevertheless, a number of insects spend much of their time under water. This is true of not a few in the perfect winged state, as for example aquatic beetles and water-bugs ('boatmen' and 'scorpions') which have some way of protecting their spiracles when submerged, and, possessing usually the power of flight, can pass on occasion from pond or stream to upper air. But it is advisable in connection with our present subject to dwell especially on some insects that remain continually under water till they are ready to undergo their final moult and attain the winged state, which they pass entirely in the air. The preparatory instars of such insects are aquatic; the adult instar is aerial. All may-flies, dragon-flies, and caddis-flies, many beetles and two-winged flies, and a few moths thus divide their life-story between the water and the air. For the present we confine attention to the Stone-flies, the May-flies, and the Dragon-flies, three well-known orders of insects respectively called by systematists the Plecoptera, the Ephemeroptera and the Odonata.

In the case of many insects that have aquatic larvae, the latter are provided with some arrangement for enabling them to reach atmospheric air through the surface-film of the water. But the larva of a stone-fly, a dragon-fly, or a may-fly is adapted more completely than these for aquatic life; it can, by means of gills of some kind, breathe the air dissolved in water.

The aquatic young of a stone-fly does not differ sufficiently in form from its parent to warrant us in calling it a larva; the life-history is like that of a cockroach, all the instars however except the final one—the winged adult or *imago*—live in the water. The young of one of our large species, a Perla for example, has well-chitinised cuticle, broad head, powerful legs, long feelers and cerci like those of the imago; its wings arise from external rudiments, which are conspicuous in the later aquatic stages. But it lives completely submerged, usually clinging or walking beneath the stones that lie in the bed of a clear stream, and examination of the ventral aspect of the thorax reveals six pairs of tufted gills, by means of which it is able to breathe the air dissolved in the water wherein it lives. At the base of the tail-feelers or cerci also, there are little tufts of thread-like gills. An insect that is continually submerged and has no contact with the upper air cannot breathe through a series of paired spiracles, and during the aquatic life-period of the stone-fly these remain closed. Nevertheless, breathing is carried on by means of the ordinary system of branching air-tubes, the trunks of which are in connection with the tufted hollow gill-filaments, through whose delicate cuticle gaseous exchange can take place, though the method of this exchange is as yet very imperfectly understood. When the stone-fly nymph is fully grown, it comes out of the water and climbs to some convenient eminence. The cuticle splits open along the back, and the imago, clothed in its new cuticle, as yet soft and flexible, creeps out. The spiracles are now open, and the stone-fly breathes atmospheric air like other flying insects. But throughout its winged life, the stone-fly bears memorials of its aquatic past in the little withered vestiges of gills that can still be distinguished beneath the thorax.

The adult dragon-fly is specialised in such a way that it captures its prey—flies and other small insects—on the wing, swooping through the air like a hawk and feeding voraciously. The head is remarkable for its large globular compound eyes, its short bristle-like feelers, and its very strong mandibles which bite up the bodies of the victims. The thorax bears the two pairs of ample wings, firm and almost glassy in texture, and its segments are projected forward ventrally, so that all six legs, which are armed with rows of sharp, slender spines, can be held in front of the mouth, where they form an effective fly-trap. The abdomen is very long and usually narrow.

A female dragon-fly after a remarkable mode of pairing, the details of which are beside our present subject, drops her eggs in the water, or lays them on water-weeds, perhaps cutting an incision where they can be the more safely lodged, or even goes down below the surface and deposits them in the mud at the bottom of a pond. From the eggs are hatched the aquatic larvae which differ in many respects from the imago. The dragon-fly larva has the same predaceous mode of life as its parent, but it is sluggish in habit, lurking for its prey at the bottom of the pond, among the mud or vegetation, which it resembles in colour. The thoracic segments have not the specialisation that they show in the imago; the abdomen is relatively shorter and broader. The larval head has, like that of the imago, short feelers, and the eyes are somewhat large, though far from attaining the size of the great globular eyes of the dragon-fly. But the third pair of jaws, forming the labium, are most remarkably modified into a 'mask,' the distal central portion (mentum) being hinged to the basal piece (sub-mentum) which is itself jointed below the head. The mentum carries at its extremity a pair of lobes with sharp fangs. Thus the mask can be folded under the head when the larva lurks in its hiding place, or be suddenly darted out so as to secure any unwary small insect that may pass close enough for capture. Dragon-fly larvae walk, and also swim by movements of the abdomen or by expelling a jet of water from the hind-gut. The walls of this terminal region of the intestine have areas lined with delicate cuticle and traversed by numerous air-tubes, so that gaseous exchange can take place between the air in the tubes and that dissolved in the water. The larvae of the larger and heavier dragon-flies (Libellulidae and Aeschnidae) breathe mostly in this way. Those of the slender and delicate 'Demoiselles' (Agrionidae) are provided with three leaf-like gill-plates at the tail, between whose delicate surfaces numerous air-tubes ramify. These gill-plates are at times used for propulsion. Thus air supply is ensured during aquatic life. But occasionally, when the water in which the larva lives is foul and poor in oxygen, the tail is thrust out of the water so that air can be admitted directly into the intestinal chamber. The aquatic life of these insects lasts for more than a year. Outward wing-rudiments are early visible on the thoracic segments; when these have become conspicuous the insect, beginning in some respects to approach the adult condition, is often called a nymph. In an advanced dragon-fly nymph, the thoracic spiracles are open, and, as the time for its final moult draws near, the insect may thrust the front part of its body out of the water, and breathe atmospheric air through these. Thus before the great change takes place the nymph has foretastes of the aerial mode of breathing which it will practise when the perfect stage shall have been attained. The emergence of the dragon-fly from its nymph-cuticle has been described by many naturalists. The nymph climbs out of the water by ascending some aquatic plant, and awaits the change so graphically sketched by Tennyson:

A hidden impulse rent the veil,
Of his old husk, from head to tail,
Came out clear plates of sapphire mail.

'From head to tail,' for the nymph-cuticle splits lengthwise down the back, and the head and thorax of the imago are freed from it, then the legs clasp the empty cuticle, and the abdomen is drawn out. After a short rest, the newly-emerged fly climbs yet higher up the water-weed, and remains for some hours with the abdomen bent concave dorsalwards, to allow space for the expansion and hardening of the wings. For some days after emergence the cuticle of the dragon-fly has a dull pale hue, as compared with the dark or brightly metallic aspect that characterises it when fully mature. The life of the imago endures but a short time compared with the long aquatic larval and nymphal stages. After some weeks, or at most a few months, the dragon-flies, having paired and laid their eggs, die before the approach of winter.

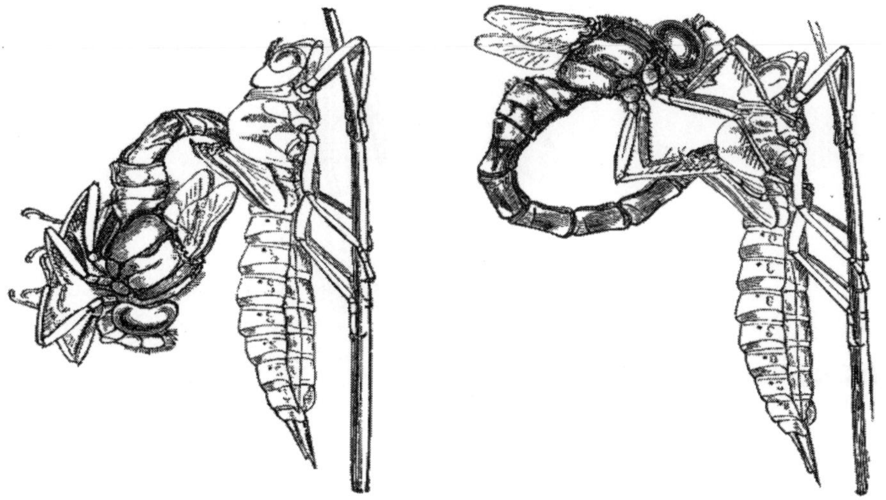

a *b*

FIG. 40 Dragon-fly (Aeschna cyanea). Two stages in emergence of fly from nymph-cuticle.

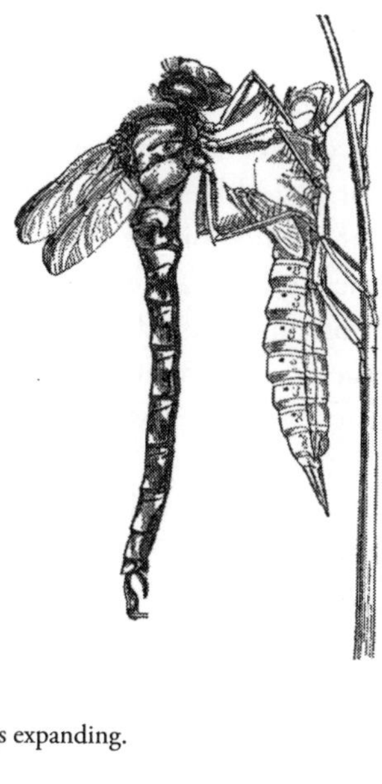

FIG. 41 Dragon-fly emerged, wings expanding.

FIG. 42 Dragon-fly (Aeschna cyanea) with expanded wings.

The life-story of a may-fly follows the same general course as that just described for the dragon-flies, but there are some suggestive differences. In the first place, we notice a wider divergence between the imago and the larva. An adult may-fly is one of the most delicate of insects; the head has elaborate compound eyes, but the feelers are very short, and the jaws are reduced to such tiny vestiges that the insect is unable to feed. Its aquatic larva is fairly robust, with a large head which is provided with well-developed jaws, as the larval and nymphal stages extend over one or two years, and the insects browse on water-weeds or devour creatures smaller and weaker than themselves. They breathe dissolved air by means of thread-like or plate-like gills traversed by branching air-tubes, somewhat resembling those of the demoiselle dragon-fly larva. But in the may-fly larva, there is a series of these gills arranged laterally in pairs on the abdominal segments, from the position and muscular attachments of these organs, for believing that they show a true correspondence to (in technical phraseology are homologous with) the thoracic legs. One feature in which the larva often agrees with the imago is the possession on the terminal abdominal segment of a pair of long jointed cerci, and in many genera a median jointed tail-process is also present, in some cases both in the larva and the imago, in others in the larva during its later stages only. The prolonged larval life in may-flies often involves a large series of moults; has enumerated twenty-one in the life-history of Chloeon. In the second year of aquatic life wing-rudiments are visible, and the larva becomes a nymph. When the time for the winged condition approaches the nymphs leave the water in large swarms. May-flies are eagerly sought as food by trout, and the rise of the fly on many lakes ushers in a welcome season to the angler.

The nymph-cuticle opens and the winged insect emerges. But this is not the final instar; may-flies are exceptional among insects in undergoing yet another moult after they have acquired wings which they can use for flight. The instar that emerges from the nymph-cuticle is a sub-imago, dull in hue, with a curious immature aspect about it. A few hours later the final moult takes place, a very delicate cuticle being shed and revealing the true imago. Then follow the dancing flight over the calm waters, the mating and egg-laying, the rapid death. The whole winged existence prepared for by the long aquatic life may be over in a single evening; at most it lasts but for a few days.

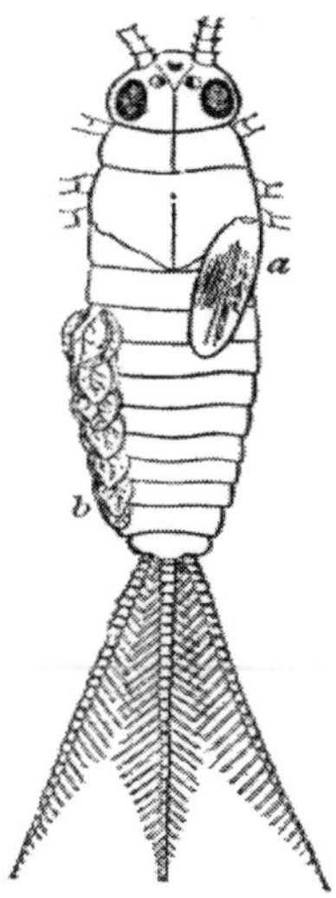

FIG. 43 Nymph of May-fly (Chloeon dipterum) showing on right side wing-rudiment (a), on left tracheal gills (b). [Feelers and legs are cut short.]

In the development of the may-flies, then, we notice not only a considerable divergence between larva and imago, both in habitat and structure; we see also what is to be observed often in more highly organised insects—a feeding stage prolonged through the years of larval and nymphal life, while the winged imago takes no food and devotes its energies through its short existence to the task of reproduction. Such division of the life-history into a long feeding, and a short breeding period has, as will be seen later, an important bearing on the question of insect transformation generally, and the dragon-flies and may-flies afford examples of two stages in its specialisation. The sub-imaginal instar of the may-fly furnishes also a noteworthy fact for comparison with other insect histories. In two points, however, the life-story of these flies with their aquatic larvae recalls that of the cockroach. All the larval and nymphal instars are active, and the wing-rudiments are outwardly visible long before the final moult.

CHAPTER FIVE
TRANSFORMATIONS – OUTWARD AND INWARD

We are now in a position to study in some detail the transformation of those insects whose life-story corresponds more or less closely with that of the butterfly, sketched in the opening pages of this little book. In the case of some of the insects reviewed in the last three chapters, the may-flies and cicads for example, a marked difference between the larva and the imago has been noticed; in others, as the coccids, we find a resting instar before the winged condition is assumed, suggesting the pupal stage in the butterfly's life-story.

The various insect orders whose members exhibit no marked divergence between larva and imago (the Orthoptera for example) are often said to undergo no transformation, to be 'Ametabola.' Those with life-stories such as the dragon-flies' are said to undergo partial transformation, and are termed 'Hemimetabola.' Moths, caddis-flies, beetles, two-winged flies, saw-flies, ants, wasps, bees, and the great majority of insects, having the same type of life-story as the butterfly, are said to undergo complete transformation and are classed as 'Metabola' or 'Holometabola.' Wherein lies the fundamental difference between these Holometabola on the one hand and the Hemimetabola and Ametabola on the other? It is not that the larva differs from the imago or that there is a passive stage in the life-history; these conditions are observable among insects with a 'partial' transformation as we have seen, though the resting instar that simulates the butterfly pupa is certainly exceptional. The most important indication of the difference between the two modes of development is furnished by the position of the wing-rudiments. In all Ametabola and Hemimetabola these are visible externally long before the penultimate instar has been reached; in the Holometabola they are not seen until the pupal stage.

Attention has already been drawn to the contrast in outward form between a butterfly and its caterpillar. As in the case of dragon-fly or may-fly, the larval period is essentially a time for feeding and growth, and during this period the larval cuticle is cast four or five, in some species even seven or eight times. After each moult some changes in detail may be observable, for example in the proportions of the body-segments or their outgrowths, in the colour or the closeness of the hairy or spiny armature. But in all main features the caterpillar retains throughout its life the characteristic form in which it left the egg. From the tiny, newly-hatched larva to the full-fed caterpillar, possibly several inches in length, there is all along the same crawling, somewhat worm-like body, destitute of any outward trace of wings. When however the last larval cuticle has split open lengthwise along the back, and has been worked off by vigorous wriggling motions of the insect, the pupa thus revealed shows the wing-rudiments conspicuous at the sides of the body, and lying neatly alongside these are to be seen the forms of feelers, legs, and maxillae of the imago prefigured in the cuticle of the pupa. The pupa thus resembles the imago much more closely than it resembles the larva; even in the proportions of the body a relative shortening is to be noticed, and the imago of any insect with complete transformation is reduced in length as compared with the full-fed larva. Now these wings and other structures characteristic of the imago, appear in the pupa which is revealed by the shedding of the last larval cuticle. From these facts we infer that the wing-rudiments must be present in the larva, hidden beneath the cuticle; and until the last larval instar, not beneath the cuticle only, but growing in such-wise that they are hidden by the epidermis. For if they were growing outwardly the new cuticle would be formed over them, so that they would be apparent after the next moult. But it is clear that only in the pupa, forming beneath the cuticle of the last larval instar, can they grow outwards.

Anatomical study of the caterpillar at various stages verifies the conclusions just drawn from superficial observation. Careful study of the imaginal discs of the wings in a caterpillar made by examining microscopically sections cut through them, shows that the epidermis is pushed in to form a little pouch (*C, p*) and that into this grows the actual wing-rudiment. Consequently the whitish disk which seems to lie within the body-wall of the larva, is really a double fold of the epidermis, the outer fold forming the pouch, the inner the actual wing-bud. Into the cavity of the latter pass branches from the air-tube system. In its earliest stage, the wing-bud is simply an ingrowing mass of cells which subsequently becomes an inpushed pouch (*B*). Until the last stage of larval life the wing-bud remains hidden in its pouch, and no cuticle is formed over it. When the pupal stage draws near the bud grows out of its sheath, and projecting from the general surface of the epidermis becomes covered with cuticle to be revealed, as we have seen, after the last larval moult, as the pupal wing. Thus all through the life of the humble, crawling caterpillar, 'it doth not yet appear what it shall be,' but there are being prepared, hidden and unseen, the wondrous organs of flight, which in due time will equip the insect for the glorious aerial existence that awaits it.

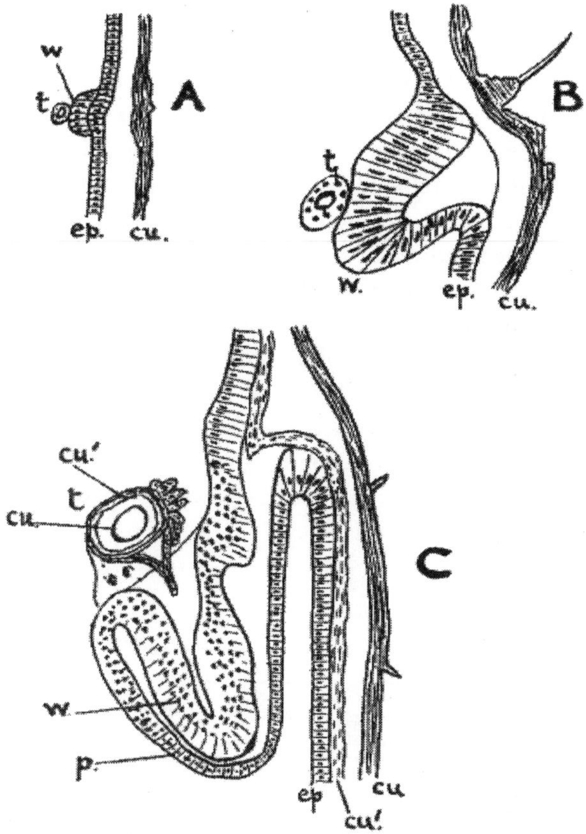

FIG. 44 Sections through epidermis and cuticle, showing three stages in growth of the imaginal disc (w) of a wing in the caterpillar of a White Butterfly (Pieris). ep, epidermis; cu, cuticle; t, air-tube, whence branches pass into the developing wing. In C, cu' represents the new cuticle forming beneath the old one, and (p) the pouch within which the wing-disc (w) lies.

As mentioned above, this hidden growth of the wing-rudiments, in butterflies, beetles, flies, bees, and the great majority of the winged insects is a character contrasting markedly with the outward and visible growth of the wing-rudiments in such insects as cockroaches, bugs, and dragon-flies. The divergence between the two modes of development is certainly very striking, and a conceivable method of transition from the one to the other is not easy to explain. Sharp has expressed the divergence by the terms *Endopterygota*, applied to all the orders of insects with hidden wing-rudiments (the 'Metabola' or 'Holometabola' of most classifications) and *Exopterygota*, including all those insects whose wing-rudiments are visible throughout growth ('Hemimetabola' and 'Ametabola'). Those curious lowly insects, belonging to the two orders of the Collembola and Thysanura, none of whose members ever develop wings at all, form a third sub-class, the *Apterygota* .

Not the wings only, but other structures of the imago, varying in extent in different orders, are formed from the imaginal discs. For example, if the thoracic leg of a late-stage caterpillar were cut off, the corresponding leg of the resulting butterfly would still be developed, although in a truncated condition. The legs of the imago are represented, through the greater part of larval life, only by small groups of cells situated within the bases of the larval legs. After the third moult these imaginal discs grow rapidly and the proximal portion of each, destined to develop into the thigh and shin of the butterfly's leg, sinks into a depression at the side of the thorax, while the tip of the shin and the five-segmented foot project into the cavity of the larval leg. Hence we understand that the amputation of the latter by the old naturalists truncated only and did not destroy the imaginal limb. In the blow-fly maggot, the imaginal discs of the legs grow out from deep dermal inpushings. Simple at first, these outgrowths by partial splitting, become differentiated into thigh and shin.

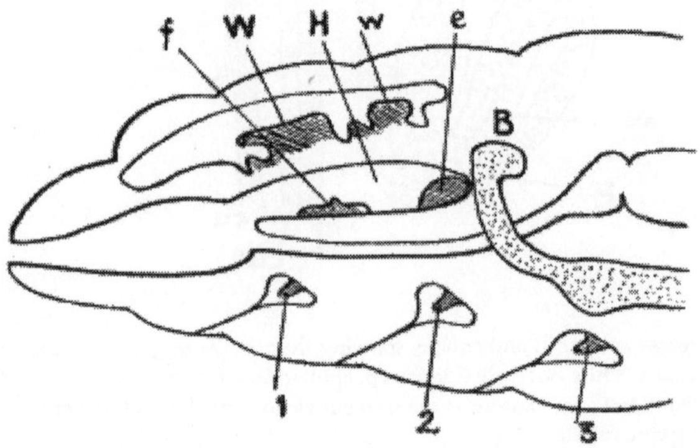

FIG. 45 Front region of Maggot of Blow-fly (Calliphora) showing diagrammatically the imaginal discs, which are shaded. e, eye; f, feeler; W, fore-wing; w, hind-wing; 1, 2, 3, legs. H is the 'cephalic vesicle,' which becomes everted at the close of the metamorphosis, so as to bring the feelers and eyes to the front, the brain (B) moving forwards at the same time.

Similarly the feelers and jaws of the butterfly are developed from imaginal discs, and this fact explains how it comes to pass that they differ so widely from the corresponding structures in the caterpillar. The larval feelers are short and stumpy, those of the butterfly long and many-jointed. The maxilla of the larva consists of a base carrying two short jointed processes; in the butterfly a certain portion of the maxilla, the hood or galea, is modified into a long, flexible grooved process, capable of forming with its fellow the trunk through which the insect sucks its liquid food. Nothing but some such provision as that of the imaginal discs could render possible the wonderful replacement of the caterpillar's jaws, biting solid food, into those of the butterfly sipping nectar from flowers.

A curious segmental displacement of the imaginal discs with regard to the larva is noticeable in some Diptera. In the larva of the harlequin-midge (Chironomus) the brain is situated in the thorax, and the imaginal discs for the head, eyes, and feelers of the adult lie in close association with it, though they arise from inpushings of the larval head. These rudiments do not appear until the last larval stage has been reached. In the gnats Culex and Corethra, on the other hand, the imaginal discs for the head-appendages retain their normal position within the larval head, and appear in an early stage of larval life. Among the flies of the bluebottle group (Muscidae) the brain is situated, as in Chironomus, in the thoracic region of the legless maggot, which is the larva of an insect of this family, and the imaginal discs for eyes and feelers lie just in front of it. Here, the imaginal buds of the legs and wings are deeply inpushed, retaining their connection with the skin only by means of a thread of cells. As the larva is legless and headless its outer form is not affected by the discs and it is not surprising to learn that they appear early. It has indeed been suggested that the pharyngeal region of the larva, in connection with which the imaginal head-discs are developed, should be regarded, though it lies in the thorax, as an inpushed anterior section of the larval head. In any case this region is pushed out during the formation of the pupa within the final larval cuticle, so that the imaginal head with its contained brain, its compound eyes, and its complex feelers, takes its rightful place at the front end of the insect.

The mention of the brain suggests a few brief remarks on the changes in the internal organs during insect transformation. There are no imaginal discs for the nervous system; the brain, nerve-cords and ganglia of the butterfly or bluebottle are the direct outcome of those of the caterpillar or maggot. Rapid but continuous changes, during the early pupal period, convert the elongate nerve-cord of the caterpillar with its relatively far-separated ganglia into the shortened, condensed nerve-cord of the Tortoise-shell butterfly (*Vanessa urticae*) with several of the ganglia coalesced. In many Diptera, on the other hand, the nervous system of the larva is more concentrated than that of the imago.

The tubular heart also of a winged insect is the directly modified survival of the larval heart.

Similarly the reproductive organs undergo a gradual, continuous development throughout an insect's life-story. Their rudiments appear in the embryo, often at a very early stage; they are recognisable in the larva, and the matured structures in the imago are the result of their slow process of growth, the details of which must be reckoned beyond the scope of this book.

On the other hand, the digestive system of insects that undergo a metamorphosis, passes through a profound crisis of dissolution and rebuilding. This is not surprising when we remember that there is often a great difference between larva and imago in the nature of the food. The digestive canal of a caterpillar runs a fairly straight course through the body and consists of a gullet, stomach (mid-gut), intestine, and rectum; it is adapted for the digestion of solid food. In the butterfly there is one outgrowth of the gullet in the head—a pharyngeal sac adapted for sucking liquids; and another outgrowth at the hinder end of the gullet (which is much longer than in the larva)—a crop or food-reservoir lying in the abdomen. The intestine of the butterfly also is longer than that of the larva, being coiled or twisted. Towards the end of the last larval stage, the cells of the inner coat (epithelium) lining the stomach begin to undergo degeneration, small replacing cells appearing between their bases and later giving rise to the more delicate epithelium that lines the mid-gut of the imago. The larval cells are shed into the cavity of the stomach and become completely broken down. The tiny replacing cells can be recognised in sections through the digestive canal of a very young larva; they may be regarded as representing imaginal buds of the adult gastric epithelium. In the transformations of two-winged flies of the bluebottle group, these replacing cells are aggregated in little masses scattered at different points along the stomach and thus corresponding rather closely to the imaginal discs of the legs and wings.

The gullet, crop, and gizzard of an insect, which lie in front of the stomach, are lined by cells derived from the outer skin (ectoderm) which is pushed in to form what is called the 'fore-gut.' Similarly the intestine and rectum, behind the stomach, are lined with ectodermal cells which arise from the inpushed 'hind-gut.' The larval fore- and hind-guts are broken down at the end of larval life and their lining is replaced by fresh tissue derived from two imaginal bands which surround the cavity of the digestive tube, one at the hinder end of the fore-gut, and the other at the front end of the hind-gut. The larval salivary glands in connection with the gullet are also broken down, and fresh glands are formed for the imago.

A large part of the substance of an insect larva consists of muscular tissue, surrounding the digestive tube, and forming the great muscles that move the various parts of the body, and of fat, surrounding the organs and serving as a store of food-material. Very many of the muscle-fibres and the fat-cells also become disintegrated during the late larval and pupal stages, and the corresponding tissues of the adult are new formations derived from special groups of imaginal cells, though some muscles may persist from the larva to the adult. Similarly the complex air-tube or tracheal system of the larva is broken down and a fresh set of tubes is developed, adapted to the altered body-form of pupa and imago.

The destruction of larval tissue and the development of replacing organs from special groups of cells, derived of course from the embryo, and carrying on the continuity of cell-lineage to the adult, are among the most remarkable facts connected with the life-story of insects. The process of tissue-destruction is known as 'histolysis'; the rebuilding process is called 'histogenesis.' In the histolysis of the two-winged flies, wandering amoeboid cells—like the white corpuscles or leucocytes of vertebrate blood—have been observed destroying the larval tissues that need to be broken down, as they destroy invading micro-organisms in the body. But students of the internal changes that accompany transformation in insects of other orders have often been unable to observe such devouring activity of these 'phagocytes,' and attribute the dissolution of the larval tissues to internal chemical changes. The fact that in all insect transformation a part, and in many a large part, of the larval organs pass over to the pupa and imago, suggests that only those structures whose work is done are broken down through the action of internally formed destructive substances, and one function of the phagocytes is to act as scavengers by devouring what has become effete and useless.

CHAPTER SIX
LARVAE AND THEIR ADAPTATIONS

Among the insects that undergo a complete transformation, there is, as we have seen in the preceding chapter, an amount of inward change, of dissolution and rebuilding of tissues, that varies in its completeness in members of different orders. It is now advisable to consider the various outward forms assumed by the larvae of these insects, or rather by a few examples chosen from a vast array of well-nigh 'infinite variety.'

In comparing the transformations of endopterygote insects of different orders, it is worthy of notice that in some cases all the members of an order have larvae remarkably constant in their main structural features, while in others there is great variety of larval form within the order. For example, the caterpillars of all Lepidoptera are fundamentally much alike, while the grubs of beetles of different families diverge widely from one another. A review of a selected series of beetle-larvae will therefore serve well to introduce this branch of the subject.

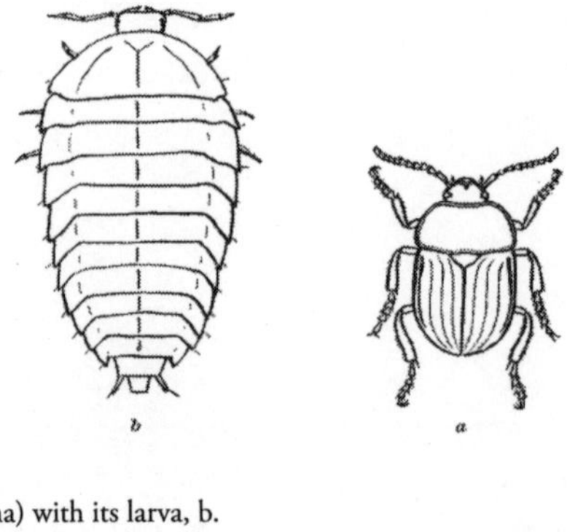

FIG. 46 Carrion-beetle (Silpha) with its larva, b.

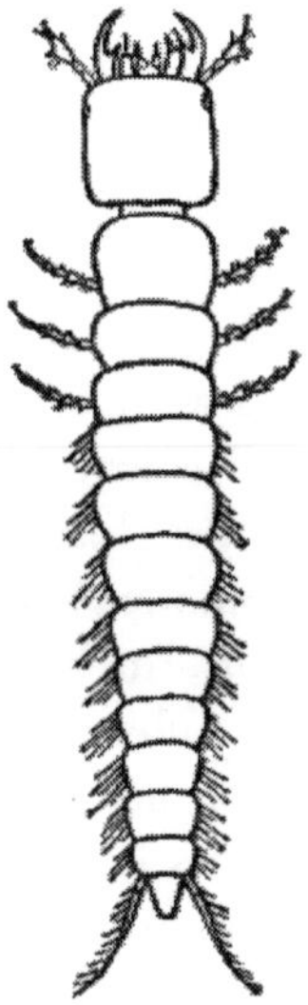

FIG. 47 Larva of a Ground-beetle (Aepus).

Beetles are as a rule remarkable among insects for the firm consistency of their chitinous cuticle, the various pieces (*sclerites*) of which are fitted together with admirable precision. In some families of beetles the larva also is furnished with a complete chitinous armour, the sclerites, both dorsal and ventral, of the successive body-segments being hard and firm, while the relatively long legs possess well-defined segments and are often spiny. Such a larva is evidently far less unlike its parent beetle than a caterpillar is unlike a butterfly. Perhaps of all beetle larvae, the woodlouse-like grub of a carrion-beetle (Silpha) or of a semi-aquatic dascillid such as Helodes shows the least amount of difference from the typical adult, on account of the conspicuous jointed feelers. The larval glow-worm, however, is of the same woodlouse-like aspect, and in this case, where the female never acquires wings, but becomes mature in a form which does not differ markedly from that of the larva, the exceptional resemblance is closer still. In all beetle-grubs the legs are simplified, there being only one segment (a combined shin and foot) below the knee-joint, whereas in the adult there is a shin followed by five, four, or at least three distinct tarsal segments. The foot of an adult beetle bears two claws at its tip, while the larval foot in the great majority of families has only one claw. In one section of the order, however, the Adephaga comprising the predaceous terrestrial and aquatic beetles, the larval foot has, like that of the adult, two claws. Some adephagous larvae, notably those of the large carnivorous water-beetles (Dyticus), often destructive to tadpoles and young fish, have completely armoured bodies as well as long jointed legs. More commonly, as with most of the well-known Ground-beetles (Carabidae), the cuticle is less consistently hard, firm sclerites segmentally arranged alternating with considerable tracts of cuticle which remain feebly chitinised and flexible. Most of the adephagous larvae have a pair of stiff processes on the ninth abdominal segment, and the insect, from its general likeness to a bristle-tail of the genus Campodea, is often called a *campodeiform* larva. From such as these, a series of forms can be traced among larvae of beetles, showing an increasing divergence from the imago. The well-known wireworms—grubs of the Click-beetles (Elateridae)—that eat the roots of farm crops, have well-armoured bodies, but their shape is elongate, cylindrical, worm-like; and their legs are relatively short, the build of the insect being adapted for rapid motion through the soil. The grubs of the Chafers (Scarabaeidae) are also root-eaters, but they are less active in their habits than the wireworms, and the cuticle of their somewhat stout bodies is, for the most part, pale and flexible; only the head and legs are hard and horny. Usually an evident correspondence can be traced between the outward form of any larva and its mode of life. For example, in the family of the Leaf-beetles (Chrysomelidae) some larvae feed openly on the foliage of trees or herbs, while others burrow into the plant tissues. The exposed larvae of the Willow-beetles have their somewhat abbreviated body segments protected by numerous spine-bearing, firm tubercles. But the grub of the 'Turnip Fly' (Phyllotreta) which feeds between the upper and lower skins of a leaf, or of *Psylliodes chrysocephala* which burrows in stalks, has a pale, soft cuticle like that of a caterpillar.

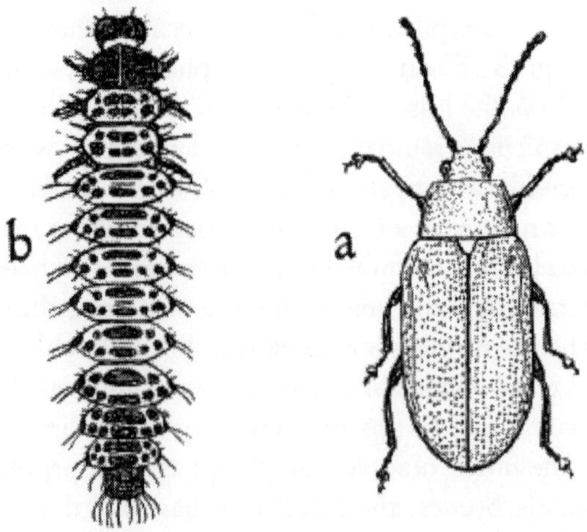

FIG. 48 (a) Willow-beetle (Phyllodecta vulgatissima) and its larva (b).

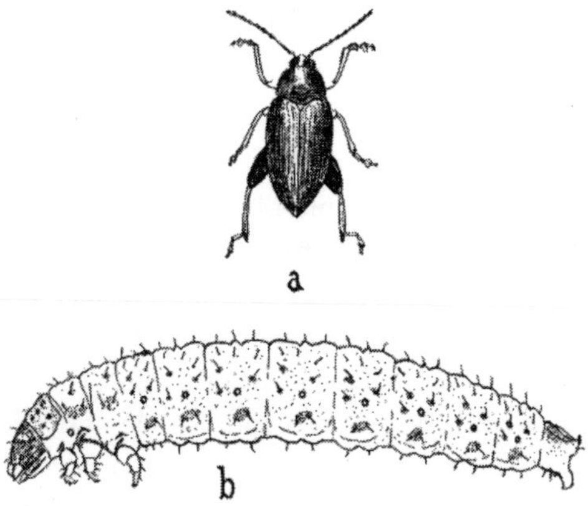

FIG. 49 (a) Cabbage-beetle (Psylliodes chrysocephala).

In the larvae of the little timber-beetles and their allies (Ptinidae), including the 'death-watches' whose tapping in old furniture is often heard, a marked shortening of the legs and reduction in the size of the head accompany the whitening and softening of the cuticle. This shortening of the legs is still more marked in the larvae of the Longhorn Beetles (Cerambycidae) burrowing in the wood of trees or felled trunks; here the legs are reduced to small vestiges.

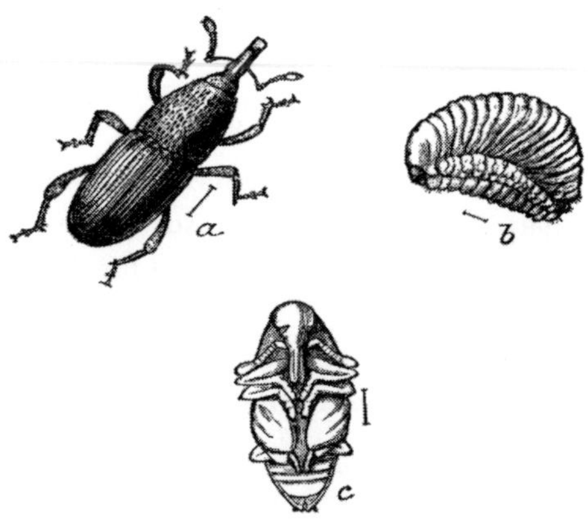

FIG. 50 a, Grain Weevil (Calandra granaria); b, larva; c, pupa.

Finally in the large family of the Weevils (Curculionidae, fig. 16) and the Bark-beetles (Scolytidae), the grubs, eating underground root or stem structures, mining in leaves or seeds, or tunnelling beneath the bark of trees, have no legs at all, the place of these limbs being indicated only by tiny tubercles on the thoracic segments. Such larvae as these latter are examples of the type called *eruciform.*

A fact of much importance in the transformations of beetles is that in a few families, the first larval instar is campodeiform, while the subsequent instars are eruciform. We may take as an example of such 'hypermetamorphosis' the life-story of the Oil or Blister-beetles (Meloidae). From the egg of one of these beetles is hatched a minute armoured larva, with long feelers, legs, and cerci, whose task is, for example, to seize hold of a bee in order that the latter may carry it, an uninvited guest, to her nest. Safely within the nest, the little 'triungulin' beetle-grub moults; the second instar has a soft cuticle and relatively shorter legs, which, as the larva, now living as a cuckoo-parasite, proceeds to gorge itself with honey, soon appear still further abbreviated. Later comes a stage during which legs are entirely wanting, the larva then resting and taking no food. The last larval instar again has short legs like the grub of the second period. In connection with this life-history we notice that the newly-hatched larva is not in the neighbourhood of its appropriate food. Hence the preliminary armoured and active instar is necessary in order to reach the feeding place; this journey accomplished, the eruciform condition is at once assumed.

In all cases indeed we may say that the particular larval form is adapted to the special conditions of life. A few examples from other orders of endopterygote insects will illustrate this point. The campodeiform type is relatively unusual, but most of the Neuroptera have larvae of this kind, active, armoured creatures with long legs, though devoid of the tail-processes often associated with similar larvae among the Coleoptera. Such are the 'Ant-lions,' larvae of the exotic lace-wing flies, which hunt small insects, digging a sandy pit for their unwary steps in the case of the best-known members of the group. The 'Aphis-lions,' larvae of Hemerobius and Chrysopa, prowl on plants infested with 'green-fly' which they impale on their sharp grooved mandibles, sucking out the victims' juices, and then, in some cases, using the dried cuticle to furnish a clothing for their own bodies. Among these insects, while the mouth of the imago is of the normal mandibulate type adapted for eating solid food, the larval mouth is constricted and the slender mandibles are grooved for the transmission of liquid food.

Turning to eruciform types of larva, we find the *caterpillar* distinguished by its elongate, usually cylindrical body with feeble cuticle, short thoracic legs and a variable number of pairs of abdominal pro-legs, universal among the moths and butterflies forming the great order Lepidoptera, and usual among the saw-flies, which belong to the Hymenoptera. The vast majority of caterpillars feed on the leaves of plants and their long worm-like bodies with the series of paired pro-legs, are excellently adapted for their habit of clinging to twigs, and crawling along shoots or the edges of leaves as they go in search of food. Of great importance to a caterpillar is its power of spinning silk, consisting of fine threads solidified from the secretion of specially modified salivary glands whose ducts open in the insect's mouth at the tip of the tubular tongue which forms a spinneret.

On the same bush caterpillars of moths and of saw-flies may often be seen feeding together. The lepidopterous caterpillar, in our countries at least, has never more than five pairs of pro-legs, situated on the third, fourth, fifth, sixth, and tenth abdominal segments; each of these pro-legs bears a number of minute hooklets, arranged in a circular or crescentic pattern, which assist the caterpillar in clinging to its food-plant. The saw-fly caterpillar, on the other hand, may have as many as eight pairs of pro-legs, the series beginning on the second abdominal segment; here, however, the pro-legs have no hooklets. Among the Lepidoptera, we notice a reduction in the number of pro-legs in the 'looper' caterpillars of Geometrid moths. Here only two pairs are present, those on the sixth and tenth abdominal segments. Consequently, as the caterpillar can cling only by the thorax and by the hinder region of the abdomen, the middle region of the body is first straightened out and then bent into an arch-like form, as the insect makes its progress by alternate movements of stretching and 'looping.'

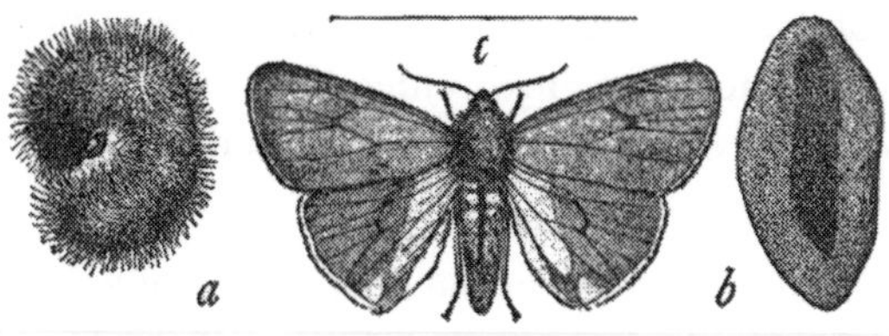

FIG. 51 c, Ruby Tiger Moth (Phragmatobia fuliginosa); a, caterpillar; b, cocoon.

Caterpillars, with their relatively soft bodies, feeding openly on the leaves of plants, are exposed to the attacks of many enemies, and the various ways in which they obtain protection are well worth studying. A clothing of hairs or spines is often present, and it is interesting to find that many species of our native Tiger and Eggar Moths (Arctiadae and Lasiocampidae) which pass the winter in the larval stage, have caterpillars with an especially dense hairy covering. Experiments have shown that hairy and spiny insects are distasteful to birds and other creatures that prey readily on smooth-skinned species, a conclusion that might well have been expected. Certain smooth caterpillars however appear to be protected by producing some nauseous secretion, which renders them unpalatable. Many of these, as the familiar cream yellow and black larva of the Magpie Moth (*Abraxas grossulariata*), are very conspicuously adorned, and furnish examples of what is known as 'warning coloration,' on the supposition that the gaudy aspect of such insects serves as an advertisement that they are not fit to eat, and that birds and other possible devourers thus learn to leave them alone. On the other hand, smooth caterpillars which are readily eaten by birds are usually 'protectively' coloured, so as to resemble their surroundings and remain hidden except to careful seekers. Many such caterpillars are green, the upper surface, which is naturally exposed to the light, being darker than the lower which is in shadow. When the caterpillar is large, the green area is often broken up by pale lines, longitudinal as on the larvae of many Owl Moths (Noctuidae) or oblique, as on the great caterpillars of most Hawk Moths (Sphingidae). Such an arrangement tends to make the insect less easily seen than were it to display a continuous area of the same colour. The 'looper' caterpillars mentioned above afford remarkable examples of 'protective' resemblance, for many of them show a marvellous likeness to the twigs of their food-plant, tubercles on the insect's body resembling closely the little outgrowths of the plant's cortex. Many caterpillars are, in their early stages, directly responsive to their surroundings as regards colour. Usually green when hatched, they remain green if kept among leaves or young shoots of plants, while they turn red, brown, or blackish if placed among twigs of these respective hues. This effect appears to be due to a direct response of the subcutaneous tissue to the rays of light reflected from the surrounding objects. The sensitiveness dies away as the caterpillar grows older, since little or no change of hue in response to a change of environment could be induced after the penultimate moult.

The 'hairs' of an insect are not in the least comparable to the hairs of mammals, being in truth, modified portions of the cuticle, secreted by special cells.

Among those families of the Lepidoptera which are usually regarded as low in the scale of organisation, caterpillars are very generally protected by the habit of feeding in some concealed situation. For example, the great larvae of the Goat Moth (Cossus) and the whitish caterpillars of the Clearwing Moths (Sesiidae) burrow through the wood of trees, eating the timber as they go. The little irritable caterpillars of the Bell Moths (Tortricidae) roll leaves, fastening the edges together with silk, and thus make for themselves a shelter; or they bore their way into seeds or fruits, like the larva of the Codling Moth that is the cause of 'worm-eaten' apples, too well-known to orchard-keepers. Very many small caterpillars mine between the two skins of a leaf, eating out the soft green tissue, and giving rise to a characteristic blister in form of a spreading patch or a narrow sinuous track through the leaf. The caterpillars of the Clothes-moths (Tineidae) make for themselves garments out of their own excrement, the particles fastened together by silk. In such curious cylindrical cases they wander over the wool or fur, feeding and indirectly supplying themselves with clothing at the same time.

The case-forming habit of the Clothes-moth caterpillars leads us naturally to consider the similar habit adopted by their allies the Caddis-larvae which live in the waters of ponds and streams, for the Caddis-flies (Trichoptera) have much in common with the more primitive Lepidoptera. The caddis-larva is as a rule of the eruciform type, but with well-developed thoracic legs, and with hook-like tail-appendages; by means of the latter it anchors itself to the extremity of its curious 'house.' It is of interest to note that in the earlier stages of some caddises, the legs are relatively very long, and the larva is quite campodeiform in aspect. Some of these caddis-grubs retain the campodeiform condition and do not shelter permanently in cases, as their relations do. Different genera of caddises differ in their mode of building. Some fasten together fragments of water-weeds and plant refuse, others take tiny particles of stone, of which they make firmly compacted walls, others again lay hold of water-snail shells, which may even contain live inhabitants, and bind these into a limy rampart behind which their bodies are in safe hiding.

The silk with which the 'caddis-worms' fasten together the materials for their houses is produced from spinning-glands which like those of the Lepidoptera open into the mouth.

The survey of the various types of beetle-larvae enumerated above concluded with a short description of the *legless grub*, which is the young form of a weevil or a bark-beetle. This is a larva in which the head alone has its cuticle firm and hard; the rest of the body is covered with a pale, flexible cuticle, so that the grub is often described as 'fleshy.' This type of larva is by no means confined to certain families of the beetles, it is frequently met with, in more or less modified form, in two other important orders of insects, the Hymenoptera and the Diptera. Among the Hymenoptera this is indeed the predominant larval type. We have just seen that a caterpillar is the usual form of larva among the saw-flies, but in all other families of the Hymenoptera we find the legless grub. A grub of this order may usually be distinguished from the larva of a weevil or other beetle, by its relatively smaller head and smoother, less wrinkled cuticle; it strikes the observer as a feebler, more helpless creature than a beetle-grub. And it is of interest to note that this somewhat degraded type of larva is remarkably constant through a great series of families—gall-flies, ichneumon-flies, wasps, bees, ants—that vary widely in the details of their structure and in their habits and mode of life. Almost without exception, however, they make in some way abundant provision for their young. The feeble, helpless, larva is in every case well sheltered and well fed; it has not to make its own way in the world, as the active armoured larva of a ground-beetle or the caterpillar of a butterfly is obliged to do.

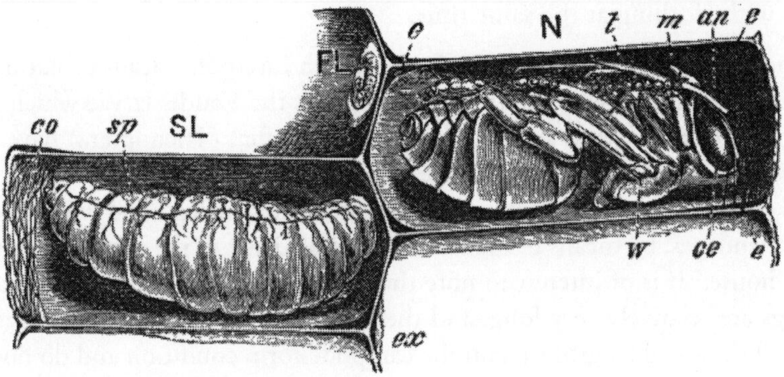

FIG. 52 Young Larva (FL), Full-grown Larva (SL) and Pupa (N) of Hive-bee (Apis mellifica). co, cocoon; sp, spiracles; ce, eye; an, feeler; m, mandible; l, labium.

Among those saw-flies whose larvae feed throughout life in a concealed situation, we find an interesting transition between the caterpillar and the legless grub. For example, the giant saw-flies (so called 'Wood-wasps') have larvae that burrow in timber, and these larvae possess relatively large heads, somewhat flattened bodies with pointed tail-end, and very greatly reduced legs. The feeble legless grub, characteristic of the remaining families of the Hymenoptera, is provided for in a well-nigh endless variety of ways. The female imago among these insects is furnished with an elaborate and beautifully formed ovipositor, and the act of egg-laying is usually in itself a provision for the offspring. Gall-flies pierce plant-tissues within which their grubs find shelter and food, the plant responding to the irritation due to the presence of the larva by forming a characteristic growth, the *gall*, pathological but often regular and shapely, in whose hollow chamber the grub lives and eats. Ichneumon-flies and their allies pierce the skin of caterpillars and other insect-larvae, laying their eggs within the victims' bodies, which their grubs proceed to devour internally. Some very small members of these families are content to lay their eggs within the eggs of larger insects, thus obtaining rich food-supply and effective protection for their tiny larvae. In Platygaster and other genera of the family Proctotrypidae, hypermetamorphosis occurs somewhat like that already described as occurring among the Oil-beetles (Meloidae). The larva of Platygaster is at first rather like a small Copepod crustacean, with prominent spiny tail-processes; after a moult this form changes into the legless grub characteristic of the Hymenoptera, among which larvae even approaching the campodeiform type are very exceptional. The species of Platygaster pass their larval stages within the larvae of gall-midges.

Wasps, bees and ants, have the ovipositor of the female modified into a sting, which is often used for the purpose of providing food for the helpless grubs. Thus the digging wasps (Sphegidae and Pompilidae) hunt for caterpillars, spiders, and other creatures which they can paralyse with their stings, and bury them alongside their eggs to furnish a food-supply for the newly-hatched young. The social wasps and many ants sting and kill flies and other insects, which they break up so as to feed their grubs within the nest. It is well known that the labour of tending the larvae in these insect societies is performed for the most part not by the mother ('Queen') but by the modified infertile females or 'workers.' Other ants and the bees feed their grubs also sheltered in well-constructed nests, on honey elaborated from nectar within their own digestive canals. In all cases we see that the helplessness of the grub is associated with some kind of parental care.

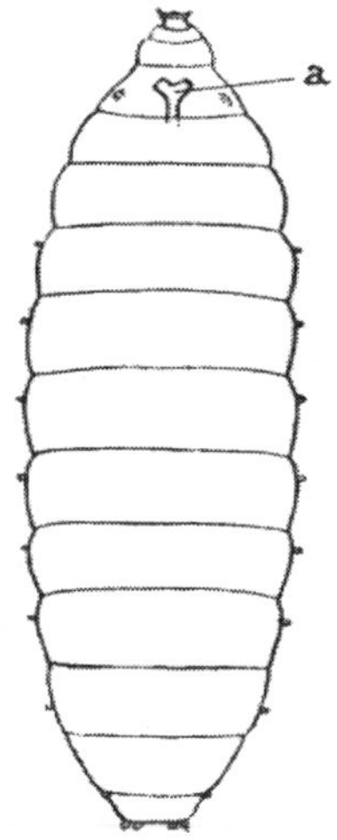

FIG. 53 Larva of Gall-midge (Contarinia nasturtii), ventral view showing anchor process (a), and spiracles projecting at sides.

From the Hymenoptera we may pass on to the Diptera or Two-winged Flies, an order of which the vast number of species and in many cases the myriads of individuals force themselves on the observer's notice. All the larvae of the Diptera are legless, those of the Crane-fly group have well-developed hard heads, with biting mandibles, but in the House-fly section the larva is of the degraded *vermiculiform* type known as the *maggot*, not only legless, but without a definite head, the front end of the creature usually tapering to the mouth, where there are a pair of strong hooks, used for tearing up the food. A few examples of each of these types must suffice in the present brief survey. A few pages back reference was made to the production of galls on various plants, through the activity of larvae of the hymenopterous family Cynipidae. Many plant-galls are due, however, to the presence of grubs of tiny dipterous insects, the Cecidomyidae or Gall-midges. A cecid grub has an elongate body with flexible, wrinkled cuticle, tapering somewhat at the two ends. The head, if rather narrow, is distinct, and beneath the prothorax is a characteristic sclerite known as the 'anchor process' or 'breast bone.' Along either side of the body is a series of paired spiracles, each usually situated at the tip of a little tubular outgrowth of the cuticle; the hindmost spiracles are often larger than the others. These little grubs live in family communities, their presence leading to some deformation of the plant that serves to shelter them. A shrivelled fruit or an arrested and swollen shoot, such as may be due respectively to the Pear-midge (*Diplosis pyrivora*) or the Osier-midge (*Rhabdophaga heterobia*), is a frequent result of the irritation set up by these little grubs. In a larva of the crane-fly family (Tipulidae) living underground and eating plant-roots, like the well-known 'leather-jacket' grubs of the large 'Daddy-long-legs' (Tipula) or burrowing into a rotting turnip or swollen fungus, like the more slender grub of a 'Winter Gnat' (Trichocera), the student notices a somewhat tough cuticle, a relatively small but distinct head, and frequently prominent finger-like processes on the tail-segment. Further examination shows a striking modification in the arrangement of the spiracles. Instead of a paired series on most of the body-segments, as in caterpillars and the vast majority of insects whether larval or adult, there are two large spiracles surrounded by the prominent tail-processes, and a pair of very small ones on the prothorax, the latter possibly closed up and useless. This restriction of the breathing-holes to a front and hind pair (amphipneustic condition) or to a hind pair only (metapneustic type) is highly characteristic of the larvae of Two-winged flies.

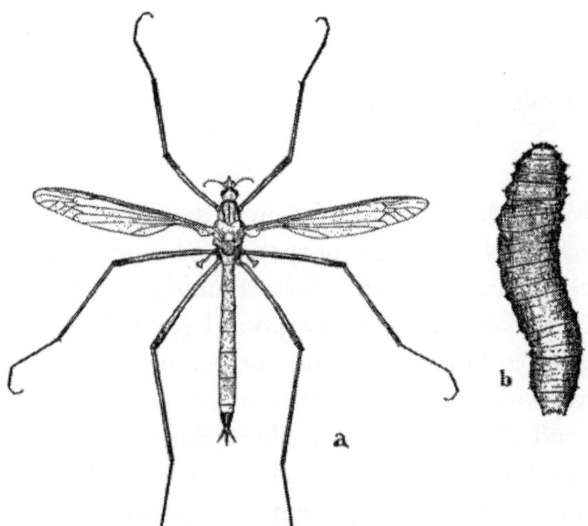

FIG. 54 Crane-fly (Tipula oleracea), a, female; b, larva ('leather-jacket' grub).

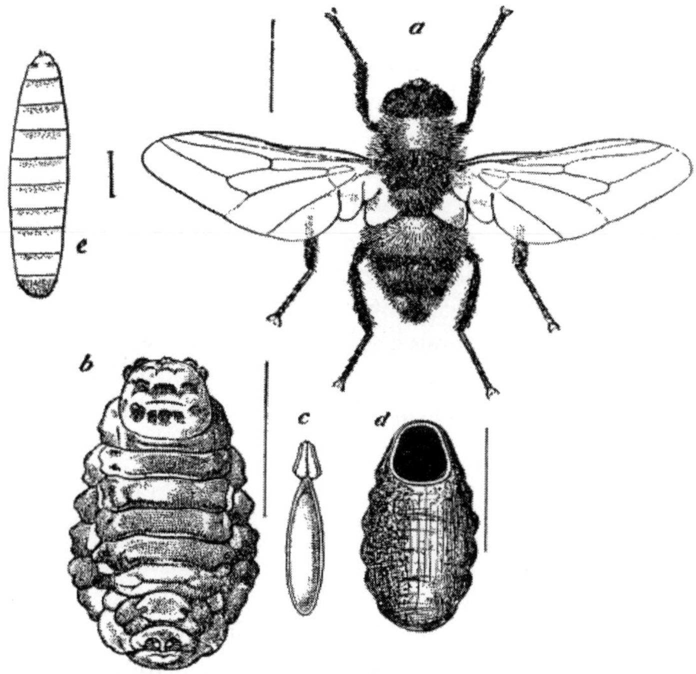

FIG. 55 Maggot of House-fly (Musca domestica), a, side-view, magnified 5 times; b, prothoracic spiracle; c, feeler; d, hind-region with posterior spiracles; e, f, head-region with mouth-hooks; g, head-region of young maggot; h, eggs.

Turning now to the *maggot,* characteristic of the House-fly section of the Diptera, we see the greatest contrast between the larva and the imago that can be found throughout the whole class of the insects. The Bluebottle's eggs, the well-known 'fly blow' laid in summer time on exposed meat, not unnaturally arouse feelings of disgust, yet they are the prelude to one of the most marvellous of all insect life-stories. The fly—with its large globular head, bearing the extensive compound eyes, the highly modified feelers with their exquisitely feathered slender sensory tips, and the complex suctorial jaws; with its compact thorax bearing the glassy fore-wings alone used for flight, though the hind-wings modified into tiny drumstick-like 'halters' are the organs of a fine equilibrating sense—is perhaps the most specialised, structurally the 'highest' of all insects. Yet in a week or two this swift, alert, winged creature is developed from the degraded maggot, white, legless, headless, that buries itself in putrid flesh, 'feeding on corruption.'

The broad end of the maggot is the tail, while the narrow extremity marks the position of the mouth. Above this are a pair of very short feelers, while from the aperture project the tips of the mouth-hooks, formidable, black, claw-like structures, articulated to the strong pharyngeal sclerites and moved by powerful muscles, tearing up the fibres of the flesh. On either side of the prothorax is an anterior spiracle, a curious branching or fan-like outgrowth, with a variable number of tiny openings which are probably of little use for the admission of air to the tubes. In many maggots the mouth-hooks and the front spiracles become more and more complex in form in the successive instars. The cuticle, white and smooth to the unaided eye, is seen on microscopic study to be set with rows of tiny spines which assist the maggot's movements through its food-mass. At the tail-end the large hind spiracles are conspicuous on a flattened dorsal area of the ninth abdominal segment; each shows a hard brown plate, traversed by three slits. And as we watch this curious degraded larva thrusting its narrow head-end into the depths of its ofttimes loathsome food-supply, we understand the advantage of access to the air-tube system being mainly confined to the hinder end of the body.

Maggots, differing from that of the Bluebottle only in minor details, are the larval forms of a vast multitude of allied species and display great variation in the nature of their food. Most, however, hide their soft defenceless bodies in some substance which affords shelter as well as food. The Bluebottle maggot burrows into flesh, that of the House-fly into horse-dung or vegetable refuse. The maggot of the Cabbage-fly eats its way into the roots of cruciferous plants, that of the Mangel-fly works out a broad blister between the two skins of a leaf, into which the newly-hatched larva crawls directly from the egg. A large number of species, forming an entire subfamily (the Tachininae) have larvae that feed as parasites within the bodies of other insects.

The habit of parasitism by maggots in back-boned animals has led to some remarkable modifications of the larva and to curious adventures in the course of the life-story. The Bot-fly of the Horse (*Gastrophilus equi*) and the Warble-fly of the Ox (*Hypoderma bovis*) lay eggs attached to the hairs of grazing animals, which, at least in the case of Gastrophilus, lick the newly-hatched larvae into their mouths. The 'bot,' or maggot of Gastrophilus, comes to rest in the horse's stomach; often a whole family attach themselves by their mouth-hooks to a small patch of the mucous coat of that organ. The maggot is relatively short and stout, with rows of strong spicules surrounding the segments, and with spiracles capable of withdrawal through a cup-like inpushing of the tail-region of the body, so that the parasite is preserved from drowning when the host drinks water. The young maggot of Hypoderma is elongate and slender, spends its first two stages burrowing in the gullet wall and then wandering through the dorsal tissues of its host; ultimately it arrives beneath the skin of the back and assumes for its third and fourth instars a broad barrel-like form. The supply of free oxygen within the ox's tissues being now insufficient, the warble-maggot bores a circular hole through the skin and rests with the tail spiracles directed upwards towards the outer air. When fully grown the maggot works its way through the hole in the host's skin, and falling to the ground pupates in some sheltered spot, the life cycle occupying about a year. Similarly the Horse-bot escapes from the host's intestine with the excrement, and pupates on the ground.

A curious modification of the maggot is noticeable in the larva of the Hover-flies (Syrphus). These, unlike most of their allies, live exposed on the foliage of plants, where they feed by preying on aphids.

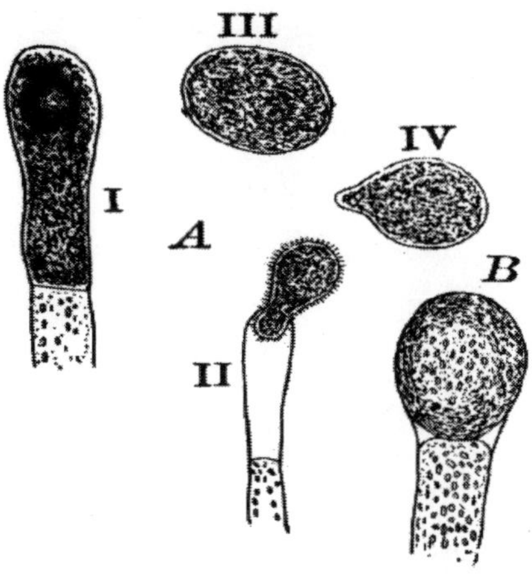

FIG. 56 Ox Warble-fly (Hypoderma bovis), a, female; b, full-grown maggot from back of ox, dorsal view; c, egg; d, empty puparium, ventral view; e, young maggot from gullet, ventral view.

In agreement with this manner of life, the cuticle is roughly granulated, often greenish or reddish in hue, and the maggot, despite its want of definite head and sense organs, moves actively and purposefully about, often rearing up on its broad tail-end with an aphid victim impaled on its mouth-hooks.

In a previous chapter reference was made to the exopterygote insects, stone-flies, dragon-flies, and may-flies, whose preparatory stages live in the water. Among the endopterygote orders many Neuroptera and Coleoptera, all Trichoptera, a very few Lepidoptera and many Diptera, have aquatic larvae. One or two examples of the adaptations of dipteran larvae to life in the water may well bring the present chapter to a close. Many members of the hover-fly family (Syrphidae) have maggots with the tail-spiracles situated at the end of a prominent tubular process. Among the best-known of syrphid flies are the drone-flies (Eristalis), often seen hovering over flowers, and presenting a curious likeness to hairy bees. The larva of Eristalis is one of the most remarkable in the whole order, the 'Rat-tailed maggot' found in the stagnant water of ditches and pools. It has a cylindrical body with the hinder end drawn out into a long telescopic tube, a more slender terminal section being capable of withdrawal into, or protrusion from, a thicker basal portion. At the extremity of the slender tube is a crown of sharp processes, forming a stellate guard to the spiracles. These processes can pierce the surface-film of the water, and place the tracheal system of the maggot in touch with the pure upper air; while its mouth may be far down, feeding among the foul refuse of the ditch, it can still reach out to the medium in which the end of its life-story must be wrought out.

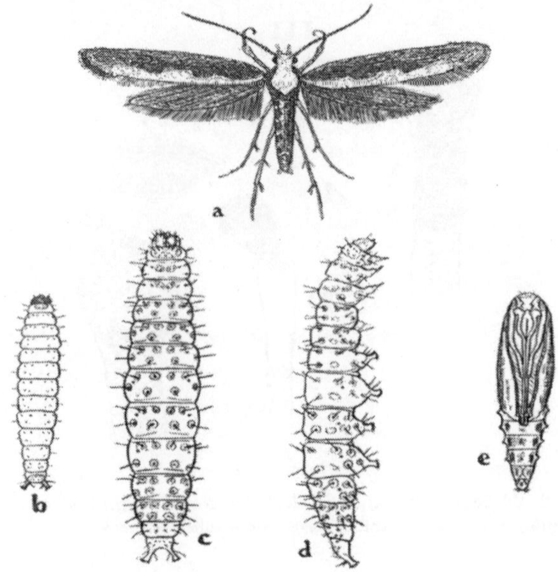

FIG. 57 Life stages of the midge (Chironomidae).

Reverting to the first great division of the Diptera, we find varied adaptations to aquatic life among many grubs that possess a definite head. The larva of a Gnat has projecting from the hind region of the abdomen a long tubular outgrowth, at the end of which are the spiracles, guarded by three pointed flaps forming a valve. When closed these pierce the surface-film of the water in which the larva lives; when opened a little cup-like depression is formed in the surface-film, from which the larva hangs. Or having accumulated a supply of air, it can disengage itself from the surface-film and dive through the water, its tracheal system safely closed. Another mode of breathing is found in the 'Blood-worms' and allied larvae of the Harlequin-midges (Chironomidae). These larvae have two pairs of cylindrical, spine-bearing pro-legs—one on the prothorax and the other on the hindmost abdominal segment; the latter structures serve to fix the larva in the muddy tube which it inhabits at the bottom of its native pond. The penultimate abdominal segment has four long hollow outgrowths, which contain blood, and have the function of gills, while the hindmost segment has four shorter outgrowths of the same nature. Enabled thus to breathe dissolved air, the Chironomus larva needs not, like the Culex or the Eristalis, to find contact with the atmosphere beyond the surface-film.

Most remarkable, in many respects, of all aquatic larvae are the grubs of the Sand-midges (Simulium). These live entirely submerged and, having no special gills, carry out an exchange of gases through the general surface of the cuticle between the dissolved air in the water and the cavities of the air-tube system. The body is shaped like a flask swollen slightly at the hinder end and possesses a median pro-leg just behind the head, also another at the tail, which serves to attach the larva to a stone or to the leaf of an aquatic plant. The head has, in addition to feelers and jaws, a pair of processes with wonderful fringes which by their motion set up currents in the water, and bring food particles within reach of the mouth. A number of the larvae usually live in a community. Their power of spinning silken threads by which they can work their way back when accidentally dislodged from their resting-place.

Examples might be multiplied, but enough have been given to enforce the conclusion that the forms of insect-larvae are wondrously varied, and that frequently, within the limits of the same order or even family, modifications of type may be found which are suited to various modes of life adopted by different insects. A survey of the multitudes of insect larvae—grubs, caterpillars, maggots—living on land, on plants, underground, in the water; feeding on leaves, in stems, on roots, on carrion, on refuse; by hunting or by lurking after prey; as parasites or as scavengers, brings home to us most strongly the conclusion that each larva is fitted to some little niche in the vast temple of life, each is specially adapted to its part in the great drama of being.

CHAPTER SEVEN
PUPAE AND THEIR MODIFICATIONS

The pupal stage is characteristic of the life-story of those insects whose larvae have wing-rudiments in the form of inpushed imaginal discs, and in all these insects there is, as we have seen, considerable divergence in form between larva and imago. In the pupa the wings and other characteristically adult structures are, for the first time, visible outwardly; it is the instar which marks the great crisis in transformation. The pupa rests, as a rule, in a quiescent condition, and during the early period of this stage the needful internal changes, the breaking down of many larval tissues, and their replacement by imaginal organs, go on. Both outwardly and inwardly therefore, the insect undergoes, at the pupal stage, a reconstruction necessitated by the differences in form and often in habit, between the larva and the winged adult; and the greater these differences, the more profound must be the changes that mark the pupal stage.

From the prominence of imaginal structures in the pupa, it is at once seen that the pupa of any insect must resemble the adult more nearly than it resembles the larva. But in different groups of insects we find different degrees of likeness between pupa and imago. In a beetle pupa, the appendages—feelers, jaws, legs, wings—stand out from the body as do those of the perfect insect. This type is called a *free* pupa. The pupal cuticle has to be shed for the emergence of the imago, but the pupa is already a somewhat reduced model of the final instar, with abbreviated wings and doubled-up legs. A free pupa is characteristic of the Coleoptera, Neuroptera, Trichoptera, Hymenoptera and many Diptera. In some cases the pupa requires to be specially adapted for a peculiar mode of life; for example, a special arrangement of breathing organs may be necessary for life under water, and there must needs be temporary pupal structures, not represented in the imago.

On the other hand, in the pupae of most Lepidoptera and of some Diptera, there is more or less coalescence between the cuticle of the appendages and the cuticle of the body generally, so that the appendages do not stand out, but being, as it were, glued down to the body, are somewhat masked. Consequently the *obtect* pupa, as this type is called, does not resemble its imago as fully as a free pupa does. The outline of the wings for example in a butterfly's pupa can in some cases be traced only with difficulty. The completely obtect pupa characterises the more highly developed families of Lepidoptera, while in the more primitive families the pupa is incompletely obtect. If the pupa of a butterfly or moth be lifted and held in the hand, a bending or wriggling motion of the abdomen can be observed. In the incompletely obtect pupa, this motion is evident in a greater number of segments than in the completely obtect, the number concerned varying from five to two in different families. In the nymphalid butterflies, the pupa is often called a 'chrysalis' on account of the golden hue displayed by the cuticle, and the term 'chrysalis' is sometimes bestowed indiscriminately on any kind of pupa. The colour of a butterfly pupa is to some extent affected by the surroundings of the caterpillar just before its last moult.

Reference has been made already to the power of spinning silk possessed by many larvae; often the principal use of this silk is to form some protection for the pupa, the larva before its last moult constructing a *cocoon* within which the pupa may rest safely. Many larvae bury themselves in the earth, and the pupa lies in an earthen chamber, the lining particles of soil fastened together by fine silken threads. Larvae that feed in wood, like the caterpillar of the Goat-moth (Cossus) make a cocoon of splinters spun together, while hairy caterpillars, such as those of the Tiger-moths, work some of their hairs in with the silk to make a firm cocoon. On the other hand, those caterpillars known as 'silkworms' make a dense cocoon of pure silk, consisting of two layers, the outer of coarse and the inner of fine threads. Silken cocoons very similar in appearance are spun by the larvae of small Ichneumon-flies. Many pupae lie in a loose cocoon formed of a few interlacing threads, as for example the conspicuous black and yellow banded pupa of the Magpie-moth (*Abraxas grossulariata*) and the pupae of various leaf-beetles. Others again spin together the edges of leaves with connecting silken threads. The grubs of bees and wasps which are reared in the comb-chambers of their nests seal up the opening of the chamber with a lid, partly silk and partly excretion, when ready to pass into the pupal state. An additional external 'capping' may be also supplied by the workers.

The pupae of butterflies are especially interesting, as illustrating the extreme reduction of the silken cocoon. The pupa of a 'swallowtail' (Papilionid) or a 'white' (Pierid) butterfly may be found attached to a twig of its food-plant or to a wall, in an upright position, its tail fastened to a pad of silk and a slender silken girdle encircling its thorax. The pupa of a 'Tortoiseshell' or 'Admiral' (Nymphalid) butterfly hangs head downwards from a twig, supported only by the tail-pad of silk, which, useless as a shelter, serves only for attachment. The pupa is fastened to this pad by a spiny hook or process, the *cremaster*. on the last abdominal segment. The cremaster is a characteristic structure in the pupa of a moth or butterfly. It corresponds with a spiny area, the suranal plate, which lies above the opening of the caterpillar's intestine. The means by which the suspended pupa of a nymphalid butterfly attaches its cremaster to the silken pad which the larva has spun in preparation for pupation, is worthy of brief attention. The caterpillar, hanging head downwards, is attached to the silken pad by its hindmost pair of pro-legs or claspers and by the suranal plate, and the cuticle is slowly worked off from before backwards, so as to expose the pupa. Were the process of moulting to be simply completed while the insect hangs by the claspers, the pupa would of course fall to the ground. But there is enough adhesion between the pupal and larval cuticles at the hinder end of the body, especially by means of the everted lining of the hind-gut, for the pupa to be supported while it jerks its cremaster out of the larval cuticle and works it into the meshes of the silken pad. The moult is thus completed and the pupa hangs securely all the time. In the numerous cases where the pupa is enclosed in a cocoon, the cremaster serves to fix the pupa to the surrounding silk. Among the more highly organised moths the pupa remains in the cocoon, the emergence being entirely left to the imago, while the pupae of the more primitive moths work their way partly out of the cocoon before the final moult begins. In the latter case, the cremaster is anchored by a strand of silk which allows a certain degree of emergence, and the pupa has rows of spines on its abdominal segments, of which a greater number retain the power of mutual motion than in those pupae which do not come out of their cocoons.

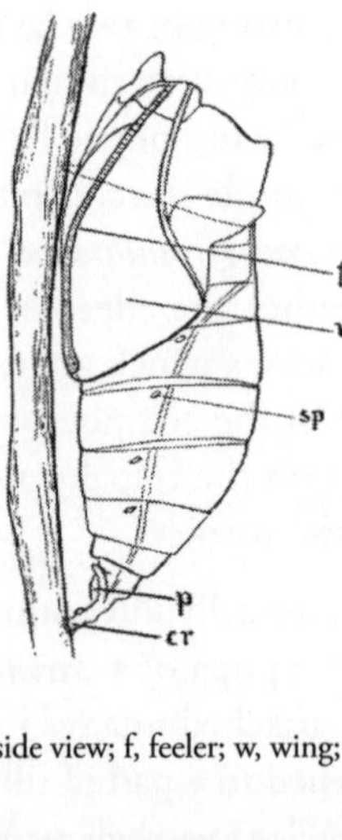

FIG. 58 Pupa of White Butterfly (Pieris), side view; f, feeler; w, wing; sp, spiracle; p, anal pro-leg; cr, cremaster.

While the pupa on the whole resembles the imago that is to emerge from it, there are not a few cases in which a special structure necessary for some contingency in pupal life is retained or adopted in this stage. A butterfly pupa, like the imago, has no mandibles, but in the case of the Caddis-flies (Trichoptera) and two families of small moths, the most primitive of all Lepidoptera, the pupa, like the larva, has well-developed mandibles. These enable the caddis pupa to bite its way out of the shortened larval case in which it has pupated, and then to swim upwards through the water ready for the caddis-fly's emergence into the air. Pupae that are submerged require special breathing-organs. In the previous chapter mention was made of the gnat's aquatic larva with its tail-spiracles adapted for procuring atmospheric air through the surface-film. The pupa of the gnat also has 'respiratory trumpets' serving the same purpose, but these are a pair of processes on the prothorax, so that the pupa, which is fairly active, hangs from the surface-film with its abdomen pointing downwards through the water. This change of position is correlated with the necessity for the imago to emerge into the air; were the pupa to hang head downwards as the larva does, the gnat would perforce have to dive into the water. With the beautifully adapted transfer of the functional spiracles, their position is appropriately arranged for the gnat's emergence at the surface, and the empty pupal cuticle floats serving the insect as a raft. On this it rests securely and the crumpled wings have opportunity to expand and harden before the insect takes to flight.

The aquatic pupae of other Diptera, many species of the midges Chironomus and Simulium for example, breathe dissolved air by means of tufts of thread-like gills, which arise on either side of the prothorax. The pupae of Simulium rest in their curious little cup-like dwellings, attached to submerged stones or plants. The Chironomus pupa is usually found in an elongate gelatinous case adhering to a stone. From this case the pupa rises to the surface of the water, that the midge may emerge into the air. When the pupal stage ends, and these gills are no longer required, their connection with the air-tube system is severed 'without undue violence.' The walls of the fine air-tubes that pass into the gills are specially strengthened, but just below the pupal cuticle these walls are exceedingly thin and delicate. Thus when the pupal cuticle is cast, they are readily broken there, and the cuticle of the midge forming beneath has a spiracular opening into the main air-trunk, ready for use during the insect's aerial life.

Among those Diptera whose larva is the headless maggot a most remarkable arrangement for protecting the pupa is to be found. The last larval cuticle, instead of being as usual worked off and cast, after separation from the underlying structures, becomes hard and firm, forming a protective case (*puparium*) within which by the processes of histolysis and histogenesis already described the organs of the pupa and imago are built up. This puparium is usually dark in colour, often brown and barrel-shaped, and a subcircular lid splits off from it at the head-end to allow the emergence of the fly. While the maggot breathes by its tail-spiracles, the functional spiracles of the puparium (connected with the tracheal system of the enclosed pupa) are far forward, and these may be situated at the tips of long sometimes branching processes, which recall the thoracic gills of the aquatic pupae mentioned a few pages above. Adaptations, various and beautiful, to special modes of life, are thus seen to characterise pupae as well as larvae.

CHAPTER EIGHT

THE LIFE-STORY AND THE SEASONS

A number of interesting questions are associated with the seasonal cycle of an insect's life-history. In a previous chapter reference has been made to the contrast between the long aquatic life of the larval dragon-fly or may-fly, extending over several years, and the short aerial existence of the winged adult restricted in the case of the may-flies to a few hours. Here we see that the feeding activities of the insect are carried on during the larval stage only; the may-fly in its winged condition takes no food, pairing and egg-laying form the whole of its appointed task. A similar though less extreme shortening of the imaginal life may be noticed in many endopterygote insects. For example, the bot- and warble-flies have the jaws so far reduced that they are unable to feed, and the parasitic life of the maggot extending over eight or nine months in the body of the horse or ox, prepares for a winged existence of probably but a few days. Again in many moths the jaws are reduced or vestigial so that no food can be taken in the winged state, as for example in the 'Eggars' (Lasiocampidae) and the 'Tussocks' (Lymantriidae). It is noteworthy that in these short-lived insects the male is often provided with elaborate sense-organs which, we may believe, assist him to find a mate with as little delay as possible; the male may-fly has especially complex eyes, while the feelers of the male silk-moth or eggar are comb-like or feathery, the branches bearing thousands of sensory hairs. A box with a captive living female of one of these moths, if taken into a wood haunted by the species becomes rapidly surrounded by a swarm of would-be suitors, attracted by the odour emitted from the prisoner's scent-glands.

Very exceptionally the imaginal stage may be omitted from the life-story altogether. The larvae of certain gall-midges (Cecidomyidae) the ovaries might become precociously mature and unfertilised eggs might be developed into small larvae observable within the body of the mother-larva; ultimately these abnormally reared young break their way out. In this case therefore there may be a series of larval generations, neither pupa nor imago being formed. Here the imaginal stage only is omitted from the life-history. Not always however is it the imaginal stage of the life-history which is shortened. Reference has already been made to the case of the virgin female aphids, whose eggs develop within the mother's body, so that active, formed young are brought forth. Among the Diptera it is not unusual to find similar cases, the female fly giving birth to young maggots instead of laying eggs. Such is the habit of the great flesh-fly (Sarcophaga), of some allied genera (Tachina, etc.) whose larvae live as parasites on other insects, and occasionally of the Sheep Bot-fly (Oestrus). In such cases we recognise the beginning of a shortened larval period; females of the dreaded African Tsetse flies (Glossinia) bring forth nearly mature larvae, which pupate soon after birth. In another group of Diptera, the blood-sucking parasites of the Hipposcidae and allied families, the whole larval development is passed through within the mother's body, and a full-grown larva is born the cuticle of which hardens and darkens immediately to form a puparium; hence these flies are often called, though incorrectly, Pupipara. Still more astonishing is the mode of reproduction in the allied family of the Termitoxeniidae, curious, degraded, wingless 'guests' of the termites, or 'white ants.' Here the individual is hermaphrodite—a most exceptional condition among insects—and lays a large egg, whence is usually hatched a fully-developed adult! Here then we find that all the early stages, usual in the higher insects, are omitted from the life-story.

Interesting comparison may be made between the total duration of various insect life-stories. To some extent at least, the length of an insect's life is correlated with its size, its food, the season of the year when it breeds. Small insects have, as a rule, shorter lives than large ones; those whose larvae devour highly nutritive food generally develop more quickly than those which have to live on dry, poor, substances; life-cycles follow one another most rapidly in summer weather when temperature is high and food plentiful.

In early chapters we have already noticed the long aquatic life of the larva and nymph of a dragon-fly, relatively a large insect, and the rapid multiplication of the repeated summer broods of virgin aphids. Within the one order of the Coleoptera it is instructive to compare the small jumping leaf-beetles, the 'turnip-flies' of the farmer, whose larvae mine in the green tissues, and complete their transformations so rapidly that several successive broods appear in the spring and early summer, with the larger click-beetles whose larvae, the equally notorious 'wireworms,' feed on roots for three or four years before they become fully grown. Among the Diptera, the 'leather-jacket' grub of the crane-fly, feeding like the wireworm on roots, has a larval life extending through the greater part of a year, while the maggot of the bluebottle, feeding on a rich meat diet, becomes mature in a few days. As examples of excessively long life-cycles the 'thirteen-year' and 'seventeen-year' cicads of North America. Certain specially populous 'broods' of these insects are known and localised, so that the appearance of the imagos in future years can be accurately predicted. Here again we have to do with bulky insects whose subterranean larvae and nymphs feed on comparatively innutritious roots.

In our own climate, it is of interest to notice the variation among insects as to the stage which carries the race over the winter. The click-beetles, mentioned just above, emerge from their buried pupae in summer, hibernate under stones or clods, and lay eggs among the herbage next spring. At the same time of course, owing to the extended term of the larval life, many more individuals of the species are wintering underground as 'wireworms' of various ages, and these, except in very severe frosts, can continue their occupation of feeding on roots. But in the case of the 'turnip-flies' the food-supply is cut off in winter, and all those beetles of the latest summer brood that survive hibernate in some sheltered spot, waiting for the return of spring, that they may lay their eggs, and start the life-cycle once again. Among the Diptera, most species pass the winter as pupae, the sheltering puparium being a good protection against most adverse conditions, or as flies. But where there is a prolonged parasitic larval life, as with the bot- and warble-flies, the maggot, warm and well-fed within the body of its mammalian host, affords an appropriate wintering stage.

Among the Hymenoptera an especially interesting seasonal life-cycle is afforded by the alternation of summer and winter generations in many Gall-flies (Cynipidae). The well-known 'oak-apples' are tenanted in summer by grubs, which after pupation develop into winged males and wingless females. The latter, after pairing, burrow underground and lay their eggs in the roots, the larvae causing the presence there of globular swellings or root-galls within which they live, pass through their transformations and develop into wingless virgin females. These shelter until February or March in their underground chambers, then climb up the tree and lay on the shoots eggs, from which will be hatched the grubs destined to grow within the oak-apples into the summer sexual brood of flies.

The Lepidoptera afford examples of hibernation in all stages of the life-history. In this order a few large moths with wood-boring caterpillars, the 'Goat' (Cossus) for example, undergo a development extending over several years, while at the other extreme a few small species may have three or more complete cycles within the twelve months. But in the vast majority of Lepidoptera we find either one or two generations, definitely seasonal, within the year; the insect is either 'single-brooded' or 'double-brooded.'

Almost every winter one or more letters may be read in some newspaper recording the writer's surprise at seeing on a sunny day during the cold season, one of our common gaily-coloured butterflies of the Vanessa group, a 'Tortoiseshell' or 'Red Admiral,' flitting about. Surprise might be greater did the observers realise that the imaginal is the normal hibernating stage for these species. Emerging from the pupa in late summer or autumn, they shelter during winter in hollow trees, under thatched eaves, in outbuildings or in similar situations, coming out in spring to lay their eggs on the leaves of their caterpillars' food-plants. The larvae feed and grow through the early summer months, in the case of the Small Tortoiseshell (*Vanessa urticae*) pupating before midsummer and developing into a July brood of butterflies whose offspring after a late summer life-cycle, hibernate; while for the larger species of the group there is, in our islands, only one complete life-cycle in the year, though the same insects in warmer countries may be double-brooded.

In September and October the pedestrian, even in a suburban square, may see moths with pretty brown, white-spotted wings flying around trees. These are males of the common 'Vapourer' (*Orgyia antiqua*), in search of the females which, wingless and helpless, rest on the cocoons surrounding the pupae whence they have just emerged, the cocoons being attached to the branches of the trees where the caterpillars have fed. After pairing, the female lays her eggs among the silk of the cocoon, partly covering them with hairs shed from her body, and then dies. The eggs thus protected remain through the winter, the larvae not being hatched till springtide, when the young leaves begin to sprout forth. The caterpillars, adorned and probably protected by their 'tussocks' of black or coloured bristles, feed vigorously. Their activity and habit of occasional migration from one tree to another, compensates, to some extent, for the females' enforced passivity; only in the larval state can moths with such wingless females extend their range. The caterpillars spin their cocoons towards the end of summer, and then pupate, the moths emerging in the autumn and the eggs, as we have seen, furnishing the winter stage.

After midsummer, the conspicuous cream, black and yellow-spotted 'Magpie' moth (*Abraxas grossulariata*) is common in gardens. The female lays her eggs on a variety of shrubby plants; gooseberry and currant bushes are often chosen. From the eggs caterpillars are hatched in autumn, but these, instead of beginning to feed, seek almost at once for rolled-up leaves, cracks in walls, crannies of bark, or similar places, which may afford winter shelters. Here they remain until the spring, when they come out to feed on the young foliage and grow rapidly into the conspicuous cream, yellow and black 'looper' caterpillars mentioned in a previous chapter. These, when fully-grown, spin among the twigs of the food-plant a light cocoon, in which the black and yellow-banded wasp-like pupa spends its short summer term before the emergence of the moth.

An equally familiar garden insect, the common 'Tiger' moth (*Arctia caia*) with its 'woolly bear' caterpillar, affords a life-cycle slightly differing from that of the 'Magpie.' The gaudy winged insects are seen in July and August, and lay their eggs on a great variety of plants. The larvae hatched from these eggs begin to feed at once, and having moulted once or twice and attained about half their full size, they rest through the winter, the dense hairy covering wherewith they are provided forming an effective protection against the cold. At the approach of spring they begin to feed again, and the fully-grown 'woolly bear' is a common object on garden paths in May and June. Before midsummer it has usually spun its yellow cocoon under some shelter on the ground and changed into a pupa.

Another modification with respect to seasonal change is shown by the Turnip moth (*Agrotis segetum*) and other allied Noctuidae (Owl-moths). These are insects with brown-coloured wings, flying after dark in June. The dull greyish larvae feed on many kinds of low-growing plants, usually hiding in the earth by day and wandering along the surface of the ground by night, biting off the farmer's ripening corn, or burrowing into his turnips or potatoes. On account of the burrowing habits of this insect it can feed throughout the winter, except when a hard frost puts a temporary stop to its activity. By April it has become fully grown and pupates in an earthen chamber a few inches below the surface. The Turnip moth in our countries is partially double-brooded, a minority of the autumn caterpillars growing more rapidly than their comrades so that they pupate, and a second brood of moths appear in September. These pair and lay eggs to reinforce the great army of wintering larvae.

Such underground caterpillars, to a great extent protected from cold, can continue to feed through the winter. With other species we find that the larva becomes fully grown in autumn, yet lives through the winter without further change. This is the case with the Codling moth (*Carpocapsa pomonella*), a well-known orchard pest, which in our countries is usually single-brooded. The moth is flying in May and lays her eggs on the shoots or leaves of apple-trees, more rarely on the fruitlets, into which however the caterpillar always bores by the upper (calyx) end. Here it feeds, growing with the growth of the fruit, feeding on the tissue around the cores, ultimately eating its way out through a lateral hole, and crawling upwards if its apple-habitation has fallen, downwards if it still remains on the bough, to shelter under a loose piece of bark where it spins its cocoon about midsummer and hibernates still in the larval condition. Not until spring is the pupal form assumed, and then it quickly passes into the imaginal state. In the south of England, and also in southwestern Ireland, this species may be double-brooded, the usual condition on the European continent and in the United States of America. There the midsummer larvae pupate at once and the moths of an August brood lay eggs on the hanging or stored fruit; in this case, again, however, the full-grown larva, quickly fed-up within the developed apples, is the wintering stage.

Several of the insects mentioned in this survey, like the last-named codling moth, are occasionally double-brooded. As an example of the many Lepidoptera, which in our islands have normally two complete life-cycles in the year, we may take the very familiar White butterflies (Pieris) of which three species are common everywhere. The appearance of the first brood of these butterflies on the wing in late April or May is hailed as a sign of advanced spring-time. They pair and lay their eggs on cabbages and other plants, and the green hairy caterpillars feed in June and July, after which the spotted pupae may be found on fences and walls, attached by the silken tail-pad and supported by the waist-girdle. In August and September butterflies of the second brood have emerged from these and are on the wing; their offspring are the autumn caterpillars which feed in some seasons as late as November, doing often serious damage to the late cruciferous crops before they pupate. The pupae may be seen during the winter months, waiting for the spring sunshine to call out the butterflies whose structures are being formed beneath the hard cuticle.

Reviewing the small selection of life-stories of various Lepidoptera just sketched, we notice an interesting and suggestive variety in the wintering stage. The vanessid butterflies hibernate as imagos; the 'vapourer' winters in the egg, the magpie as a young ungrown larva, the 'tiger' as a half-size larva; the Agrotis caterpillar feeds through the winter, growing all the time; the codling caterpillar completes its growth in the autumn, and winters as a full-size resting larva; lastly, the 'whites' hibernate in the pupal state. And in every case it is noteworthy that the form or habit of the wintering stage is well adapted for enduring cold.

Our native 'whites' afford illustration of another interesting feature often to be noticed in the life-story of double-brooded Lepidoptera. The butterflies of the spring brood differ slightly but constantly from their summer offspring, affording examples of what is called *seasonal dimorphism*. All three species have whitish wings marked with black spots, larger and more numerous in the female than in the male. In the spring butterflies these spots tend towards reduction or replacement by grey, while in the summer insects they are more strongly defined, and the ground colour of the wings varies towards yellowish. In the 'Green-veined' white (*Pieris napi*) the characteristic greenish-grey lines of scaling beneath the wings along the nervures, are much broader and more strongly marked in the spring than in the summer generation, whose members are distinguished by systematic entomologists under the varietal name *napaeae.* In tropical countries where instead of an alternation of winter and summer, alternate dry and rainy seasons prevail, somewhat similar seasonal dimorphism has been observed among many butterflies. Not a few forms of Precis, an African and Indian genus that had long been considered distinct species are now known to be alternating seasonal forms of the same insect. The offspring when adult does not closely resemble the parent; its appearance is modified by the climatic environment of the pupa.

We are thus led to see from the life-story of such insects, that the course of the story is not rigidly fixed; the creature in its various stages is plastic, open to influence from its surroundings, capable of marked change in the course of generations. And so the seasonal changes in the history of the individual from egg to imago point us to changes in the age-long history of the race.

CHAPTER NINE

PAST AND PRESENT: THE MEANING OF THE STORY

In the previous chapter we recognised how the seasonal changes in various species of butterflies as observable in two or three generations, indicate changes in the history of the race as it might be traced through innumerable generations. The endless variety in the form and habits of insect-larvae and their adaptations to various modes of life, which have been briefly sketched in this little book, suggest vaster changes in the class of insects, as a whole, through the long periods of geological time. Every student of life, influenced by the teaching of Charles Darwin and his successors, now regards all groups of animals from the evolutionary standpoint, and believes that comparisons of facts of structure and life-history of orders and classes evidently akin to each other, furnish at least some indications of the course of development in the greater systematic divisions, even as the facts of seasonal dimorphism, mentioned in the last chapter, give hints as to the course of development in those restricted groups that we call species or varieties. A brief discussion of the main outlines of the life-story of insects in the wide, evolutionary sense may thus fitly conclude this book.

In the first place we turn to the 'records' of those rocks, in whose stratified layers are entombed remains, often fragmentary and obscure, of the insects of past ages of the earth's history. Compared with the thousands of extinct types of hard-shelled marine animals, such as the Mollusca, fossil insects are few, as could only be expected, seeing that insects are terrestrial and aerial creatures with slight chance of preservation in sediments formed under water.

We are now considering insects from the standpoint of their life-histories, and the individual life-story of an insect of which we possess but a few fragments of wings or body, entombed in a rock formed possibly before the period of the

Coal Measures, can only be a matter of inference. Still it may safely be inferred that when the structure of these remains clearly indicates affinity to some existing order or family, the life-history of the extinct creature must have resembled, on the whole, that of its nearest living allies. And all the fossil insects known can be either referred to existing orders, or shown to indicate definite relationship to some existing group.

Passing over some doubtful remains of Silurian age, we find in rocks usually regarded as Devonian the most ancient fossils that can be certainly referred to the insects, while from beds of the succeeding Carboniferous period, a number of insect remains have been disinterred. These Palaeozoic insects were frequently of large size, and they show distinct affinities with our recent may-flies, dragon-flies, stone-flies, and cockroaches. In the Permian period, the latest of the divisions of the Palaeozoic, lived Eugereon, an insect with hemipteroid jaws and orthopteroid wings. All these insects must have been exopterygote in their life-history, if we may trust the indications of affinity furnished by their structure. In the Mesozoic period, however, insects with complete transformations must have been fairly abundant. Rocks of Triassic age have yielded beetles and lace-wing-flies, while from among Jurassic fossils specimens have been described as representing most of our existing orders, including Lepidoptera, Hymenoptera and Diptera. In Cainozoic rocks fossil insects of nearly six thousand species have been found, which are easily referable to existing families and often to existing genera. We may conclude then, imperfect though our knowledge of extinct insects is, that some of the most complex of insect life-stories were being worked out before the dawn of the Cainozoic era. Some instructive hints as to differences in the rate of change among different insect groups may be drawn from the study of parasites. For example, an identical species of the Mallophaga (Bird-lice) infests an Australian Cassowary and two of the South American Rheas; while two species of the same genus (Lipeurus) are common to the African Ostrich and a third kind of South American Rhea. These parasites must have been inherited unchanged by the various members of these three families of flightless birds from their common ancestors, that is from early Cainozoic times at latest. On the other hand, the various kinds of such highly specialised parasites as the warble-flies of the oxen and deer, must have become differentiated during those later stages of the Cainozoic period which witnessed the evolution of their respective mammalian hosts.

The foregoing brief outline of our knowledge of the geological succession of insects shows that the exopterygote preceded, in time, the endopterygote type of life-history. We have already seen that those insects undergoing little change in the life-cycle, and with visible, external wing-rudiments, are on the whole less specialised in structure than those which pass through a complete transformation. These two considerations, taken together, suggest strongly that in the evolution of the insect class, the simpler life-history preceded the more complex. Such a conclusion seems reasonable and what might have been expected, but we are confronted with the difficulty that if the most highly organised insects pass through the most profound transformations, then insects present a remarkable and puzzling exception to the general rules of development among animals. A few students of insect transformation have indeed supposed that the crawling caterpillar or maggot must be regarded as a larval stage which recalls the worm-like nature of the supposed far-off ancestors of insects generally. However, the caterpillar or the maggot is itself a specialised product of the evolutionary process, adapted to its own particular mode of larval life.

The explanation of insect transformation is, in brief, to be found in an increasing amount of divergence between larva and imago. The most profound metamorphosis is but a special type of growth, accompanied by successive castings and renewings of the chitinous cuticle, which envelopes all arthropods. In the simplest type of insect life-story, there is no marked difference in form between the newly-hatched young and the adult, and in such cases we find that the young insect lives in the same way as the adult, has the same surroundings, eats the same food. This is the rule (see Chapters II and III) with the Apterygota, the Orthoptera, and most of the Hemiptera. In the last-named order, however, we find in certain families marked divergence between larva and imago, for example in the cicads, whose larvae live underground, while in the coccids, whose males are highly specialised and females degraded, there succeeds to the larva—very like the young stage in allied families—a resting instar, which in the case of the male, suggests comparison with the pupa of a moth or beetle.

Turning to the stone-flies, dragon-flies and may-flies, whose life-stories have been sketched in Chapter IV, we find that the early stages are passed in water, whence before the final moult, the insects emerge to the upper air. Except for the possession of tufted gills, adapting them to an aquatic life, the stone-fly nymphs differ but slightly from the adults; the grubs of the dragon-flies and may-flies, however, are markedly different from their parents. In connection with these comparisons, it is to be noted that the dragon-flies and may-flies are more highly specialised insects than stone-flies, divergent specialisation of the adult and larva is therefore well illustrated in these groups, which nevertheless have, like the Hemiptera and Orthoptera, visible external wing-rudiments.

From the vast array of insects that show internal wing-growth and a true pupal stage, a few larval types were chosen for description in Chapter VI, and a review of these suggests again the thought of increasing divergence between larva and imago. Reference has been made previously to the many instances in which the former has become pre-eminently the feeding, and the latter the breeding stage in the life-cycle. It seems impossible to avoid the conclusion that the active, armoured campodeiform grub differing less from its parent than an eruciform larva differs from its parent, is as a larval type more primitive than the caterpillar or maggot.

A certain amount of difficulty may be felt with regard to the theory of divergent evolution between imago and larva, in the case of those insects with complete transformation whose grubs and adults live in much the same conditions. By turning over stones the naturalist may find ground-beetles in company with the larvae of their own species. On the leaves of a willow tree he may observe leaf-beetles (Phyllodecta and Galerucella) together with their grubs, all greedily eating the foliage; or lady-bird beetles (Coccinella) and their larvae hunting and devouring the 'greenfly.' All of these insects are, however, Coleoptera, and the adult insects of this order are much more disposed to walk and crawl and less disposed to fly than other endopterygote insects. Their heavily armoured bodies and their firm shield-like forewings render them less aerial than other insects; in many genera the power of flight has been altogether lost. It is not surprising, therefore, that many beetles, even when adult, should live as their larvae do; since the acquirement of complete metamorphosis they have become modified towards the larval condition, and an extreme case of such modification is afforded by the wingless grub-like female Glow-worm (Lampyris).

With most insects, however, the larva must be regarded as the more specially modified, even if degraded, stage. The insect grub is not a precociously hatched embryo, like the larvae of multitudes of marine animals, but that it exhibits in a modified form the essential characters of the adult. Comparison for example can be readily made between the parts of the caterpillar and the butterfly, whose story was sketched in the first chapter of this book, widely different though caterpillar and butterfly may appear at a superficial glance. And the survey of variety in form, food, and habit of insect larvae given in Chapter VI enforces surely the conclusion that the larva is eminently plastic, adaptable, capable of changing so as to suit the most diverse surroundings. The larva must be regarded as the more modified stage, because while all the adult's structures are represented in the larva, even if only as imaginal buds, there are commonly present in the larva special adaptive organs not found in the imago, for example the pro-legs of caterpillars or the skin-gills of midge-grubs. The correspondence of parts in butterfly and caterpillar just referred to, may still be traced, though less easily, in bluebottle and maggot. The latter is an extreme example of degenerative evolution, and its contrast with the elaborately organised two-winged fly marks the greatest divergence observable between the larva and imago. With this divergence the resting pupal stage, during which more or less dissolution and reconstruction of organs goes on, becomes a necessity, and it has already been pointed out how the amount of this reconstruction is greatest where the divergence between the larval and perfect stages is most marked. Whatever differences of opinion may prevail on points of detail, the general explanation of insect metamorphosis as the result of divergent evolution in the two active stages of the life-story must assuredly be accepted. No other explanation accords with the increasing degree of divergence to be observed as we pass from the lower to the higher insect orders.

The successive incidents of the life-story of most insects are largely connected with the acquisition of wings. Wings, and the power of flight wherewith they endow their possessors, are evidently beneficial to the race in giving power of extending the range during the breeding period and thus ensuring a wide distribution of the eggs. In no case are wings fully developed until the closing stage of the insect's life, they are always acquired after hatching or birth. We have already noticed the essential difference between the exopterygote and endopterygote insects, the wing-rudiments of the former growing outwards throughout life while those of the latter remain hidden until the pupal instar. Reference has already been made to insects of various orders in which one sex is wingless, the Vapourer Moth for example, or all the individuals of both sexes are wingless, as the aberrant cockroaches mentioned in Chapter II, or certain generations of virgin females are wingless, for example aphids and gall-flies. Insects may thus become secondarily wingless, that is to say be manifestly the offspring of winged parents, and such wingless forms may on the other hand give rise to offspring or descendants with well-developed wings. Frequently, as in the case of the aphids, many wingless generations intervene between two winged generations. A striking illustration of this fact is afforded by an aquatic bug, *Velia currens*, commonly to be seen skating over the surface of running water. The adults of Velia are nearly always wingless, but now and then the naturalist meets with a specimen provided with functional wings, the possession of which enables the insect to make its way to a fresh stream. Moreover there are whole orders of parasitic insects, such as the lice and fleas, which, showing clear affinity to orders of winged insects, are believed to be secondarily wingless. And from the analogy of the periodic loss and recovery of wings in various generations of the same species, he has concluded that the gap between the exopterygote and the endopterygote method of development may have been bridged by an anapterygote condition; that the ancestors of those insects with complete transformations were the wingless descendants of primitive insects which grew their wings from visible external rudiments, and that in later times re-acquiring wings, they developed these organs in a new way, from inwardly directed rudiments or imaginal buds.

On the whole the most probable suggestion which can be made as to the origin of 'complete' transformation in insects is that the instar in which wings were first visible externally became later and later in the course of the evolution of the more highly organised groups. In this way a gradual transition from the exopterygote to the endopterygote type of life-story is at least conceivable. It will be remembered that a may-fly undergoes a moult after acquiring functional wings, emerging into the air as a 'sub-imago.' In not a few endopterygote insects, the pupa shows more or less activity, swimming through water intermittently (gnats) or just before the imago has to emerge (caddis-flies); working its way out of the ground (crane-flies) or coming half-way out of its cocoon (many moths). The pupa of the higher insects almost certainly corresponds with the may-fly's sub-imago, and the facts just recalled as to remnants of pupal activity suggest that in the ancestors of endopterygote insects what is now the pupal instar was represented by an active nymphal or sub-imaginal stage, possibly indeed by more than one stage, as Packard and other writers have stated that pupae of bees and wasps undergo two or three moults before the final exposure of the imago. Such an early pupal instar has been defined as a 'pro-nymph' or a 'semi-pupa.' Examples have been given of the exceptional passive condition of the penultimate instar in Exopterygota. The instars preceding this presumably had originally outward wing-rudiments in all insect life-histories, and the endopterygote condition was attained by the postponement of the outward appearance of these to successively later stages. The leg and wing rudiments of the male coccid beneath the cuticle of the second instar are strictly comparable to imaginal buds, and these are present in one instar of what is generally regarded as an exopterygote life-history. The first instar in all insects has no visible wing-rudiments, but when they grow outwardly from the body, they necessarily become covered with cuticle, so that they must be visible after the first moult. There is no supreme difficulty in supposing that the important change was for these early rudiments to become sunk into the body, so that the cuticle of the second, and, later, of the third and succeeding instars, showed no outward sign of their presence. Whatever may be the exact explanation of these abnormalities, they show that in the life-story of the higher insects outward wing-rudiments may even yet appear before the pupal stage, confirming our belief that such appearance is an ancestral character. The inward growth of these wing-rudiments may well have been correlated with a difference in form between the newly-hatched insect and its parent. As this difference persisted until a constantly later stage, and the pre-imaginal instar became necessarily a stage for reconstruction, the present condition of complete metamorphosis in the more highly organised orders was finally attained.

To explain satisfactorily these complex life-stories is however admittedly a difficult task. The acquisition of wings is, as we have seen, a dominating feature in them all, but if we try to go yet a step farther back and speculate on the origin of wings in the most primitive exopterygote insects, the task becomes still more difficult. It is indeed highly probable that insects had a common origin with aquatic Crustacea, but all the evidence points to the ancestors of insects having become breathers of atmospheric air before they acquired wings. How the wings arose, what function their precursors performed before they became capable of supporting flight, we can hardly even guess.

Our study of the life-story of insects, therefore, while it has taught us something of what is going on around us to-day, and has given us hints of the course of a few threads of that long life-story which runs through the ages, brings us face to face with the most instructive, if humbling fact that 'there are many more things of which we are ignorant.' The passage from creeping to flight, as the caterpillar becomes transformed into the butterfly, was a mystery to those who first observed it, and many of its aspects remain mysterious still. Perhaps the most striking result of the study of insect transformation is the appreciation of the divergent specialisation of larva and imago, and it is a suggestive thought that of the two the larva has in many cases diverged the more from the typical condition. The caterpillar crawling over the leaf, or the fly-grub swimming through the water, may thus be regarded as a creature preparing for a change to the true conditions of its life. It is a strange irony that the preparation is often far longer than the brief hours of achievement. But the light which research has thrown on the nature of these wonderful life-stories, the demonstration of the unseen presence and growth within the insect, during its time of preparation among strange surroundings, of the organs required for service in the coming life amid its native air, confirm surely the intuition of the old-time students, who saw in these changes, so familiar and yet so wonderful, a parable and a prophecy of the higher nature of man.

OUTLINE CLASSIFICATION OF INSECTS

Class INSECTA or HEXAPODA.

Sub-class A, Apterygota.

Order

Thysanura (Bristle-tails).
Collembola (Spring-tails).

Sub-class B, Exopterygota.

Order

Dermaptera (Earwigs).
Orthoptera (Cockroaches, Grasshoppers, Crickets).
Plecoptera (Stone-flies).
Isoptera (Termites or 'White Ants').
Corrodentia
Copeognatha (Book-lice).
Mallophaga (Biting-lice).
Ephemeroptera (May-flies).
Odonata (Dragon-flies).
Thysanoptera (Thrips).
Hemiptera
Heteroptera (Bugs, Pond-skaters)
Homoptera (Cicads, 'Greenfly,' Scales).
Anoplura (Lice).

Sub-class C, Endopterygota.

Order

Neuroptera (Alder-flies, Ant-lions, Lacewings).
Coleoptera (Beetles).
Mecaptera (Scorpion-flies).
Trichoptera (Caddis-flies).
Lepidoptera (Moths and Butterflies).
Diptera (Two-winged flies)
Orthorrhapha (Crane-flies, Midges, Gnats)
Cyclorrhapha (Hover-flies, House-flies, Bot-flies, &c).
Siphonaptera (Fleas).
Hymenoptera
Symphyta (Saw-flies)
Apocrita (Gall-flies, Ichneumon-flies, Wasps, Bees, Ants).

TABLE OF GEOLOGICAL SYSTEMS

These names, given by geologists to the various divisions of rocks, as indicated by the fossils entombed in them, are arranged in 'descending' order, the more recent formations above, the more ancient below, as newer deposits necessarily lie over older beds.

Calnozoic or Tertiary Group.

Pleistocene.

Pliocene.

Miocene.

Eocene.

Mesozoic or Secondary Group.

Cretaceous.

Jurassic.

Triassic.

Palaeozoic or Primary Group.

Permian.

Carboniferous.

Devonian.

Silurian.

Cambrian.

GLOSSARY OF ENTOMOLOGICAL TERMS

A

A: prefix, is privative; wanting or without.

Ab: off; away from.

Abbreviated: cut short; not of usual length.

Abdomen: the third or posterior division of the insect body: consists normally of nine or ten apparent segments, but actual number is a mooted question: bears no functional legs in the adult stage.

Abdominal: belonging or pertaining to the abdomen.

Abdominal feet: see pro-legs.

Abdominal groove: the concave lobe of the inner margin of secondaries enveloping the abdomen beneath, in some butterflies.

Abdominal pouch: in female Parnassiids, a sac-like ventral cavity, formed by material secreted during copulation.

Abductor: applied to muscles that open out or extend an appendage or draw it away from the body: see adductor.

Abductor mandibulae: the muscle that opens the mandibles.

Aberrant: unusual; out of the ordinary course.

Aberration: a form that departs in some striking way from the normal type; either single or occurring rarely, at irregular intervals.

Abiogenesis: spontaneous generation.

Abnormal: outside the usual range or course; not normal.

Aborted: a structure developed so as to be unfit for its normal function obsolete or atrophied.

Abraded: scraped or rubbed.

Abrupt: suddenly or without gradation.

Abscissus: cut off squarely, with a straight margin.

Absconditus: hidden, concealed; retracted into another.

Acalyptrata: those muscid flies in which alulae are absent or rudimentary.

Acanthus: a spine, spur or prickle.

Acaudal -ate: without a tail.

Accessory: added, or in addition to.

Accessory carinae: in Orthoptera the lateral carinae of the face.

Accessory cell: a cell not commonly present in the group; in some orders of definite location as, e.g. in Lepidoptera, usually a small cell at the end of the subcosta, giving rise directly or indirectly to veins 7 to 10:= 1st radius 2 (Comst.); = areole.

Accessory glands: any glands opening into the ducts of the reproductive system.

Accessory sac: a glandular structure of the female reproductive system containing a sticky secretion.

Accessory subcostal vein: the vein given off from the subcosta and branching toward the apex of the wing in Perlidae.

Aceous or aceus: suffix; similar to, or of the nature of.

Acephalous: without a head.

Acerata: arthropods without true antennae Arachnids and Limulus

Acetabular caps: Hemiptera; the coxal cavity.

Acetabuliform: like a shallow saucer with more or less incurved sides.

Acetabulum: the cavity into which an appendage is articulated; specifically the coxal cavity, - q.v.; also applied to a cup-like cavity in the sucking mouth of maggots.

Achreioptera: ordinal term proposed for the coleopterous family Platypsyllidae.

Achromatic: free from color; tissue that does not stain readily.

Acicular: needle-shaped; with a long, slender point.

Aciculate: a surface that appears as if scratched with a needle.

Acidotheca: the pupal sheath of the ovipositor.

Acini: granulations, like those on a blackberry: the terminal secreting tubes of glands.

Acinose -ous: a surface set with acini.

Acone: applied to compound eyes in which the individual ocelli have no crystalline cone or lens; see eucone.

Acoustic nerve: connects the auditory pits or other organs of hearing with special ganglia.

Acridophagus: preying and feeding on grasshoppers.

Acrostichal bristles: Diptera; two rows of bristles on the middle of the dorsum; specifically, minute peculiar bristles on the dorso-central region of Dolichopodidae.

Aculeata: Hymenoptera; the stingers, including bees and wasps.

Aculeate: prickly; armed with short, sharp spines; specifically, in Hymenoptera furnished with a sting which is a modified ovipositor and connected with a poison sac.

Aculeus -ei: a prickle; a small sharp point; specifically, an ovipositor, especially when sting-like, as in Hymenoptera; in male Tipulidae a slender, horny, often curved and pointed piece, projected when the forceps is open.

Acuminate: tapering to a long point.

Acupunctate: a surface with fine punctures as if made with a needle.

Acutangulate: forming, or meeting in an acute angle.

Acute: pointed: terminating in or forming less than a right angle.

Acutilingual: with a sharp pointed tongue or mouth structure, as in some bees.

Acutilingues: bees with a short pointed tongue: see obtusilingues.

Addorsal: close to but not quite on the middle of the dorsum.

Addorsal line: in caterpillars, is longitudinal, a little to one side of the dorsal and between it and the subdorsal line.

Adductor: applied to muscles that draw an appendage to the body or bring parts into apposition: see abductor.

Adductor mandibulae: the muscle that draws in or closes the mandible.

Adeloceratous: with concealed antennae: see cryptocerata.

Adephagous: belonging to the Adephaga: pentamerous, predatory, terrestrial beetles with filiform antennae and predatory habits: see hydradephagous.

Adherent: attached or clinging to.

Adipose: fat or fatty: see fat-body.

Adiscota: insects that develop into adults without forming imaginal discs; see discota.

Adminicula: supports or props: the spinous processes on the abdomen of boring and burrowing pupae.

Adnate: adjoining; adhering or growing together: closely connected.

Adpressed: laid or pressed to; contiguous.

Adsperse -us: with markings of closely crowded small spots.

Adsternal: situated next or close to the sternum.

Adult: the stage when an insect is sexually mature and ready to reproduce normally.

Aduncate -cus, -catus: a part gradually bent through its whole extent.

Adventitious: occurring accidentally, out of the ordinary course, without apparent reason.

Adventral line: in caterpillars, extends along the under side between the middle and the base of legs.

Adventral tubercle: on the abdominal segments of caterpillars on the inner base of the leg, and correspondingly on the apodal segments; constant: is number VIII of the abdominal series (Dyar).

Aeneous -eus: shining bronze or brassy.

Aenescent: becoming or appearing bronzed or brassy.

Aequale: equal.

Aequilate-us: of equal breadth throughout.

Aerial: living in the air; applied to flying insects.

Aeriductus: a spiracle: the tracheal, gill-like structures of aquatic larvae: more specifically the tail-like extensions of rat-tailed maggots and some aquatic Hemiptera.

Aeroscepsin: an indefinite sense of perception supposed to be located in the antenna.

Aeroscepsy: The faculty of observing atmospheric changes: supposed to be located in the antenna.

Aerostats: a pair of large air sacs at base of abdomen in Diptera.

Aeruginose -us: the color of verdigris [blue green].

Aestival: occurring in summer.

Aestivation: applied to summer dormancy.

Afferent: carrying inwardly or toward the centre.

Affinis: related to: similar in structure or development.

Afternose: a triangular piece below antennae and above clypeus: see post-clypeus.

Agamic -ous: reproducing without union with a male.

Agamogenesis: reproduction without fertilization by a male: see parthenogenesis; gamogenesis.

Agglomerate: heaped or massed together.

Agglutinate: stuck or glued together; welded into one mass.

Aggregated: crowded together as closely as possible.

Agnathous: without jaws; specifically applied to those Neuropteroid series in which the mouth structures are obsolescent.

Aileron: the scale covering the base of primaries in some insects; see tegulae in Diptera = alula and squama, q.v.

Air-sacs or vesicles: pouch-like expansions of tracheal tubes in heavy insects, capable of inflation and supposed to lessen specific gravity.

Air-tube: a respiratory siphon.

Ala -ae: a wing or wings.

Alar appendage: see alulet.

Alar frenum: a small ligament crossing the supra-alar groove toward the root of the wing: Hymenoptera.

Alary: relating to the wings: applied also to the wing muscles of heart.

Alate -us: winged; with lobes similar to wings in appearance though not necessarily in function.

Albi, albus: white.

Albicans: formed or made of white.

Albidus: white with dusky tinge.

Albinic: of the character of an albino.

Albinism: that condition in which there is an absence of color or a whitening in a form usually colored.

Albino: a colorless individual of a species that is normally colored.

Albumen: the white of egg or the substances in the tissues which have the same characteristics.

Albumin: the characteristic substance forming the white of egg.

Albuminoid: like or of the character of albumen.

Alimentary canal: the digestive tract as a whole; begins at the mouth and extends through the body to the anus.

Alitrunk: that part of the thorax to which the wings are attached: in many Hymenoptera, includes the 1st abdominal segment.

Alizarine: a transparent, orange red [alizar crimson].

Alligate -us: fastened or suspended by a thread; like the chrysalis of Papilio, etc.

Alliogenesis: when the development includes an alternation of generations (q.v.), as in Cynipids.

Alluring glands: glandular structures diffusing an odor supposed to be attractive to the opposite sex.

Allux: next to the last joint of tarsus; in Rhynchophora.

Alpine zone: = arctic zone, q.v.

Alternation of generations: where a species that occurs in both sexes periodically produces only parthenogenetic females; the latter, in turn, producing the sexed form; occurs in Cynipidae and some Homoptera: see heterogeny.

Altus: above: applied to a part raised above the usual level.

Alulae: Diptera; a pair of membranous scales above the halteres, behind the root of the wing, one above or before the other; the anterior attached to the wing and moving with it, the posterior fastened to the thorax and stationary; see calyptra; squama; squamula; lobulus; axillary lobe; aileron; scale; tegulae: Coleoptera; a membranous appendage of the elytra which prevents dislocation.

Alulet: Diptera: the lobe at basal posterior part of wing; = alar appendage; posterior lobe: and has been used as = alula.

Alutaceous: rather pale leather brown [burnt sienna]: covered with minute cracks, like the human skin.

Alveolate: furnished with cells: deeply pitted.

Alveolus: a cell, like that of a honeycomb.

Amber: a transparent, clear, pale yellowish brown; of the color of amber [a mixture of pale cadmium yellow and a little burnt umber].

Ambient vein: Diptera; the costal vein when it extends beyond the apex and practically margins the wing.

Ambrosia: bee-bread: the food cultures of certain Scolytid beetles.

Ambulatoria: that series of Orthoptera in which the legs are fitted for walking only; Phasmids.

Ambulatorial: fitted for walking or making progress on the surface.

Ambulatorial setae: specialized hairs or bristles, situated on the ventral segments of the abdomen of some Coleoptera. Ambulatory: moves by walking; formed for walking.

Ametabola -ous: insects without obvious metamorphoses, in which the larvae usually resemble the adult and the pupae are active.

Ametabolion: an insect that has no distinct metamorphoses.

Amethystine -us: bright blue with a reddish admixture; clear like an amethyst [between mauve and lilac].

Amnion: the inner of the two membranes enveloping the embryo.

Amnion cavity: a tube-like insinking from the ventral plate of the embryo, extending cephalad.

Amnion fold: the extensions of the amnion which close the mouth of the amnion cavity in the embryo.

Amnios: the first cast skin of the larva when a moult occurs almost immediately after emergence from the egg.

Amoebiform: having the appearance or properties of an amoeba.

Amoeboid: applied to movements similar to those of an amoeba.

Amphibiotica: those pseudoneuropterous insects whose larvae are aquatic but whose imagos are aerial; stone-flies; May-flies; dragon- flies.

Amphimixis: the mingling of the germ plasm of two individuals.

Amphiodont: applied to those forms of male Lucanids bearing mandibles of medium size, between teleodont and priodont; =mesodont.

Amphipneustic: applied to larvae which have the spiracles confined to the anterior and terminal segments.

Ample: broad; large; sufficient in size.

Amplected: when the head is received into a concavity of the prothorax; e.g. Hister.

Ampliate -us: moderately dilated.

Amplificatus: dilated; enlarged.

Ampulla: Orthoptera; an extensile sac between head and prothorax used by the young in escaping from oötheca, and later, in molting: Heteroptera; a blister-like enlargement at the middle of the anterior margin of the pro-thorax.

Ampulla-like: flask-shaped; applied to a vascular sac at base of antennae which aids in the blood circulation of head and its appendages.

Amygdaliform: almond-shaped.

Anabolic: the constructive change from food material to animal tissue: see katabolic.

Anal: pertaining or attached to the last segment of the abdomen; the point or angle of any wing or other appendage that is near to or at any time reaches the tip of the abdomen.

Anal angle: on the secondaries is that angle nearest the end of the abdomen when the wings are expanded: the angle between the inner and outer margin of any wing; = hind angle of primaries.

Anal appendages: generally; applied to the external genital parts.

Anal area: Orthoptera and Neuroptera; the hinder or anal portion of a wing within the anal vein = axillary area.

Anal cells: the spaces between the anal veins (Comst.): in Diptera, anal cell (Will.), the space nearest the body, inclosed by the 5th and 6th veins sometimes called the third basal cell (Coq.) = 1st anal (Comst.).

Anal field: Orthoptera; that area on the tegmina corresponding to the anal area of the secondaries.

Anal filaments: see caudal setae.

Anal fork: applied to the cerci of Coleopterous larvae.

Anal foot: applied to the tip of the body in larval Chironomids, which is modified to serve as a hold-fast.

Anal furrow: in wings, lies between the cubitus and 1st anal vein.

Anal glands: appendages of the alimentary canal, opening into it near the posterior extremity, secreting either a lubricant, a silk-gum, or some other specialized material.

Anal horns: in Collembola, are small processes borne on the last abdominal segment.

Anal lobes: in Lecaniinae, a pair of small, triangular, hinged processes forming a valve which covers the anal orifice.

Anal loop: Odonata; the loop formed by the angulations of 1st anal vein.

Analogous: similar in function; but differing in origin and structure: e.g. the wings of birds and insects: see homologous.

Anal operculum: the dorsal arch of the 10th abdominal segment; in caterpillars = supra-anal plate, q.v.

Anal organs: Collembola; the two modified hairs arising from a tubercle ventro-cephalad of the anus and usually curving caudo-dorsad.

Anal orifice: see anus.

Anal papilla: Collembola; see anal tubercle.

Anal plate: in caterpillars, the shield-like covering of the dorsum of the last segment: in the embryonic larva the 11th tergite.

Anal ring: a chitinous ring encircling the anus in many Coccidae.

Anal scale: one of the lateral processes of the ovipositor in Cynipidae, lying outside and below the lateral scale.

Anal siphon: the anal breathing tube of Culicid larvae.

Anal style: a slender process on or within the terminal segment of the abdomen in Homoptera.

Anal tubercle: Collembola; the tubercle bearing the anal organs: = anal papilla.

Anal tubercles: a pair of prominent, rounded or conical processes, situate one on each side of the anus in certain Coccids.

Anal valves: see podical plates.

Anal veins: those longitudinal unbranched veins extending from base to outer margin below the cubitus; the first anal, also termed vena dividens, q.v., is the 6th of the series starting from the base, and it may be followed by several others which are numbered in order to the inner margin.

Anastomosing: inosculating or running into each other.

Anastomosis: a running together; usually applied to wing veins, often to markings; sometimes used like stigma, q.v.; also in Neuroptera, a series of cross-veinlets nearly in one row; a connecting series of veinlets.

Anceps: two-edged; similar to ensiform, q.v.

Ancestral: primitive; inherited from an earlier form or ancestor.

Anchor process: = breastbone, q.v.

Anchylosed: grown together at a joint.

Ancipital: with two opposite edges or angles.

Androconia: specialized, usually small scales of peculiar form, found localized on some male butterflies.

Androgynous: uniting the characters of both sexes.

Aneurose: a wing without veins except near costa.

Angle: of tegmina, "is the longitudinal ridge formed along the interno-median by the sudden flexure from the horizontal to the vertical portion when closed."

Angular area: Hym.; the posterior of the three areas on the metanotum between the lateral and pleural carinae; = 3rd pleural area.

Angulate: forming an angle; when two margins meet in an angle.

Angulose: having angles.

Angulus: forming an angle: = angulate.

Angustatus: narrowed; narrowly drawn out.

Anisoptera: that division of the Odonata in which the hind wings are wider, especially at base, than the front wings.

Annectent: applied to connecting or intermediate forms.

Annelet or annellus: Hym.; small ring-joints between scape and funicle.

Annulate: ringed or marked with colored bands.

Annulet: a small or narrow ring or annulus.

Annuliform: in the form of rings or segments.

Annulus: a ring encircling a joint, segment, spot or mark; sometimes applied to the inner ring encircling the mouth opening.

Annulus antennalis: the ring sclerite of the head into which the basal segment of the antennae is inserted; = antennal sclerite.

Anomalous: unusual; departing widely from the usual type.

Anoplura: wingless species without metamorphosis, habits epizoötic, thoracic segments similarly developed: a composite aggregation which includes both the biting and sucking lice.

Ante: before; used as a prefix.

Ante-alar sinus: Odonata; a grooved area extending transversely immediately in front of the base of each front wing.

Ante-apical: just before the apex.

Ante-clypeus: Odonata; the lower of the two divisions of the clypeus; the inferior half of the clypeus whenever there is any apparent line of demarcation: = clypeus-anterior; infra-clypeus; rhinarium; second clypeus.

Ante-coxal piece: Coleoptera; that portion of the metasternum lying in front of the posterior coxae, often passing between them and meeting the abdomen of mandible, is the lateral sclerite of the clypeus; - one on each side.

Ante-cubital: see ante-nodal, cross veins and spaces.

Ante-furca: an internal forked process from the prosternum, to which muscles are attached.

Ante-humeral: relating to the space just before origin of wings.

Ante-humeral stripe: Odonata; a discolored stripe, approximately parallel to, but to the inner side of the humeral suture, q.v.

Antemedial line: = t. a. line, q.v.

Antemedian: Diptera; applied to leg-bristles situated before the middle.

Antenna -ae: two jointed, sensory organs, borne, one on each side of the head, commonly termed horns or feelers.

Antenna-cleaner: a fringed excavation on the interior base of the 1st segment of the anterior tarsi of Hymenoptera which, when covered by the movable process from the end of the tibia, forms an opening through which the antennae may be drawn: similar structures are on the fore tibiae of Carabid beetles: tarsal claws are also used by various insects to clean antennae.

Antennal appendage: in Mallophaga, a projecting process of the 1st or 3rd segment in the male.

Antennal formula: in Coccidae; made by enumerating the antennal joints in the order of their length, beginning with the longest and bracketing together those of the same length.

Antennal fossa -w: grooves or cavities in which antennae are located or concealed: = a. grooves: antennary fossa.

Antennal fovea: Diptera; a groove or grooves in the middle of the face as though for the lodgment of the antennae; bounded on the sides by the facial ridges.

Antennal foveolae: Orthoptera; the pits between frontal costa and lateral carinae, in which the antennae are inserted.

Antennal grooves: see antennal fossa.

Antennal lobes: of brain, see deuto-cerebrum.

Antennal organs: in Collembola are sensory structures on the distal segment.

Antennal process: Diptera; the frontal protuberance upon which the antennae are inserted.

Antennal sclerite: see annulus antennalis.

Antennal segment: the second or deutocerebral segment of head.

Antennary fossa: see antennal fossa.

Antennary furrow: in Mallophaga, grooves on the under side of the head in which the antennae lie.

Antenniferous: bearing antennae.

Antenniform: made up like, or having the appearance of antennae.

Antennule: a small antennae or feeler-like process.

Antenodal cells: Odonata; in Agrionidae the cells included between the short sector (M 4 Comst.) and the upper sector of the triangle (Cu 1, Comst.), and between the quadrilateral (or quadrangle) and the vein descending from the nodus.

Antenodal cross veins: Odonata; extend between costa and subcosta, and between subcosta and media, from the base to the nodus, forming the antenodal or ante-cubital cells: = ante-cubital.

Antenodal costal spaces: Odonata; the cells between costa and subcosta, from the base to the nodus: = ante-cubitals.

Anteocular: the region just before the eye; specifically applied in Collembola to a peculiar structure of undefined function situated in front of the eyes: = prostemmatic.

Antepectus: the lower surface of the prothorax.

Antepenultimate: the last but two.

Anterior: in front; before; in Dip., that face of the leg which is visible from the front when the leg is laterally extended and bristles on that face are anterior.

Anterior branch of third vein, in Diptera (Will.), = radius 4 (Comst.).

Anterior field: Orthoptera; of tegmina, see costal field.

Anterior intercalary vein: Diptera; = media 2 (Comst.); of Loew = discoidal vein.

Anterior lamina: Odonata; the anterior sternal border of abdominal segment 2, modified to form the front margin of the genital pocket.

Anterior lobe: Orthoptera; see lobes.

Anterior squama: = antisquama; q.v.

Anterior stigmatal tubercle: on thoracic and abdominal segment of caterpillars; varies from substigmatal to stigmatal anterior; sometimes united to IV: it is V of the abdominal series, IV of the thorax (Dyar).

Anterior trapezoidal tubercle: on thoracic and abdominal segment of caterpillars addorsal, anterior, always present, rarely united with II: it is I of the abdominal series, la of the thorax (Dyar).

Antero: to the front; anteriorly.

Antero-dorsal: Diptera; applied to leg bristles at the meeting of anterior and dorsal face.

Antero-ventral: Diptera; applied to leg bristles at the meeting of anterior and ventral face.

Anthobian: feeding on flowers; applied to certain lamellicorn Coleoptera in which the labium extends beyond the mentum.

Anthophila: Hymenoptera; species in which the basal joint of the hind tarsus is dilated and pubescent; the bees.

Anthracine -us: coal black; black with a bluish tinge.

Anti: over against; opposite; contrary: (prefix).

Anticus: frontal; belonging to or directed toward the front.

Antigeny: opposition or antagonism of the sexes; embracing all forms of secondary sexual diversity.

Antipodal costal spaces: Odonata; the cells between costa and subcosta, from the base to the modus; = antecubitals.

Antisquama: Diptera; the upper of the two which moves with the wings; = antitegula; see also squama.

Antitegula: see antisquama.

Antlia: the spiral tongue or haustellum of Lepidoptera.

Antliata: insects with a sucking mouth; originally applied to Lepidoptera and Diptera, later and more specifically to Diptera.

Antrorse -sum: directed toward the front.

Anus: the end of the digestive tract, through which the food remnants are passed: the posterior part of the individual: specifically, in Coccidae, a more or less circular opening on the dorsal surface of the pygidium, varying in location as regards the circumgenital gland orifices: = anal orifice.

Aorta: the anterior, narrow part of the heart, opening into the head.

Apex: that part of any joint or segment opposite the base by which it is attached; that point of a wing furthest removed from base or at the end of the costal area.

Aphaniptera: indistinctly winged; see Siphonaptera.

Aphideine: see aphidilutein.

Aphidilutein: a yellowish fluid found in plant lice, changed to a rich violet by alkaline reagents.

Apical: at, near or pertaining to the apex; usually of a wing.

Apical area: see petiolar area.

Apical areas: apical cells in some Homoptera.

Apical cell: a cell near or at the apex of a wing; in Hymenoptera (Norton) = medial (Comst.); outer apical cell = 2d medial 2 (Comst.); inner apical cell = medial 3 (Comst.).

Apical cells or cellules: Trichoptera; the series of cells along the outer margin of wing from pterostigma to arculus.

Apically: toward or directed toward the apex.

Apical sector: one of the longitudinal veins in the apical part of wing of Neuroptera.

Apical transverse carina: Hymenoptera; crosses the metanotum behind middle and separates the median from the posterior cells or areas.

Apiculis: an erect, fleshy short point.

Apiculate: covered with fleshy, short points.

Apivorous: devouring bees.

Apneustic: without an open tracheal system; respiration is through the skin or through tracheal gills.

Apocrita: = petiolate, q.v.

Apodal: with single, simple tubercles instead of feet, in larvae; without feet = apodous.

Apode: one that has no feet.

Apodema: a conspicuous transverse band crossing the thorax in front of the scutellum in male Coccidae.

Apodeme: an inwardly directed process to which a muscle is attached.

Apodous: without feet; see apodal.

Apolar: without differentiated poles; without apparent radiating processes applied to cells.

Apophysis: the lower of the two joints of trochanter in ditrocha trochanterellus; the dorso-lateral metathoracic spines in Hymenoptera; also used as synonymous with ento-thorax.

Apophystegal plates: Orthoptera; flattened blade or plate-like sclerites covering the gonapophyses.

Apotypes: = hypotypes; q.v.

Appendage -es: any part, piece or organ attached by a joint to the body or to any other main structure.

Appendice -es: any attached body or small process; an appendix.

Appendicial: supplementary: relating to appendices.

Appendicle: a small appendix: in some bees, a small sclerite at tip of labrum.

Appendiculate: bearing appendages; said of antennae where the joints have articulated appendages; of tarsal claws that have membranous processes at base.

Appendiculate cell: Hymenoptera; is on costa just beyond 2d radius 1 and 2.

Appendigerous: bearing appendages.

Appendix: a supplementary or additional piece or part, added to or attached to another: in Heteroptera; = cuneus, q.v.

Appress -ed: to press against; closely applied to.

Approximate: near to; applies to antennae inserted close together.

Aptera: those that have no wings: an ordinal term formerly employed for fleas, lice and other wingless forms now distributed in other orders: later used for the simplest or lowest insects, including the Thysanura and Collembola.

Apterodicera: wingless, with two antennae.

Apterous: without wings.

Apterygogenea: those insects that are wingless in all stages and presumed to be descended from ancestors which never were winged: see pterygogenea.

Apterygota: = apterygogenea; see pterygote.

Aquamarine -us: sea green: pale green with predominant blue and a little gray [nile green].

Aquatic: living wholly in water.

Aquatilia: cryptocerous Hemiptera of truly aquatic habit.

Arachnoideous: resembling or similar to a cobweb.

Araneiform: spider-like in appearance.

Arboreal: living in, on, or among trees.

Arborescent: branching like the twigs of a tree.

Archaic: ancient; no longer dominant; of the olden time.

Archiptera: those Neuroptera with incomplete metamorphosis = Pseudo-neuroptera.

Arctic Zone: is that part of the boreal region above the limit of tree growth restricted to the area above timber line on the summits of high mountains: = alpine.

Arcuate: curved like a bow: = arcuate.

Arcuato-emarginate: with a bow-like or curved excision.

Arculus: Odonata; a small cross vein between radius and cubitus near the base, leaving an elongate triangle between them: Trichoptera; a point, often hyaline, on the forewing where the cubitus (or post cubitus) runs into the margin: in Homoptera; a cross-veinlet nearly reaching posterior margin at same point as in Trichoptera: in other orders applied to a cross-vein in similar position, apparently giving rise to the median.

Arcus: a bow; part of a circle; but less than one half.

Area mediastinal, scapularis and ulnaris: the areas in front of the mediastinal, the scapular, and the ulnar veins in Orthoptera.

Areae or Areolae: wing cells or spaces between veins.

Arenicolous: applied to species frequenting sandy areas.

Arenose: a surface that is sandy or gritty.

Areola: a small cell on the wings of certain Hemiptera: see also areae Hymenoptera; the central of three median areas on the metanotum: = 2d median area; upper median area.

Areolate: with small defined areas, like a network.

Areole: Lepidoptera; see accessory cell, cell and cellule.

Areolet: one of the small spaces between veins of net-veined insects.

Argentate: shining, silvery white.

Argenteous: silvery.

Argillaceous: of the texture, appearance or color of clay.

Arid: Applied to regions in which the normal rainfall is insufficient to produce ordinary farm crops without irrigation, and in which desert conditions prevail: see humid.

Arista: a specialized bristle or process on antennae of certain Diptera.

Aristate: Diptera; that type of antennae that bears an arista: = athericerous.

Aristiform: of the form or appearance of an arista.

Armature: applied to the spinous or chitinous processes on the legs, body or wings; or the corneous parts of genitalic structures.

Armatus: set with spines, claws or other chitinous processes.

Armillate: with a ring or annulus of raised or different tissue.

Arolium -ia: cushion-like pads on the tarsi of many insects: one of the lobes of the pulvillus; in Orthoptera, used only for the terminal pad between the claws: see empodium; pulvillus; palmula; plantula; onychium, paronychium, pseudonychium.

Arquate: see arcuate.

Arrhenotokous: capable of producing male offspring only, as in worker bees and some saw-flies.

Arrhenotoky: parthenogenetic reproduction when the progeny are all males: see thelyotoky and deuterotoky.

Arthrium: Coleoptera; the minute, concealed tarsal joint in pseudotetramera and trimera.

Arthroderm: the outer skin or covering of articulates.

Arthrodial: an articulation that permits motion in any direction.

Arthromere: a body segment or ring: = somite.

Arthropleure: the side piece of an arthromere.

Arthropods: all those articulates having jointed legs.

Article: a joint or segment.

Articular pan: the cup or dish-like depression forming the socket into which an articulation is fitted.

Articulate: that branch of the animal kingdom whose members are made up of rings, segments or articulations.

Articulate: divided into joints or segments.

Articulated apex: see clasp filament.

Articulation: the point or place where two parts or segments are joined: also applied to an individual joint or segment.

Articulatory epideme: the partly chitinized membrane by which the wings are attached to the thorax.

Artus: the organs of locomotion generally.

Asexual: applied where the reproductive organs are incompletely developed and eggs or young are produced by cell-budding: = parthenogenetic.

Ash-gray: a mixture of black and white, with a faint orange tinge: like ashes of anthracite coal.

Aspect: indicates the direction to which a surface faces or in which it is viewed; it may be dorsal, ventral, caudal, cephalic or lateral.

Asperities: surface roughenings or dot-like elevations.

Aspersus: rugged, with distinct elevated dots.

Assembling: gathering together; applied when a virgin female is exposed to attract such males as may be near, either to secure a pairing or merely to obtain specimens; also called sembling.

Assurgent: down-curved at base, then upcurved to an erect position.

Asymmetrical: not alike on the two sides; not symmetrical.

Asymmetry: a state of unlikeness in lateral development; absence of symmetry in form or in the development of members.

Ater: deep black; not shining.

Aterimus: the deepest black.

Athericerous: see aristate.

Atom -us: a minute dot or point.

Atomarius: with minute dots or points.

Atrachelia: Coleoptera in which there is no visible constriction between head and prothorax: Rhynchophora and some Heteromera.

Atrium: a chamber just within the spiracle and before the occluding structure to the trachea.

Atrocoeruleus: very deep, blackish, sky-blue.

Atrophied: wasted away; unfit for use.

Atropurpureus: dark purplish, nearly black [an admixture of mauve and black].

Atrous: jet black.

Atrovelutinus: velvety black.

Atrovirens: dark green, approaching blackish [prussian green].

Attenuated: drawn out; slender; tapering.

Attingent: touching.

Atus: suffix; denotes possession of a quality or structure.

Atypic -ical: off type; not of the usual form.

Auchenorhynchus: with the beak issuing from the inferior portion of head, as in Homoptera.

Auditory: relating to the sense of hearing.

Auditory organs: Orthoptera; specialized structures covered by a tense membrane, on the anterior tibia or base of abdomen; any structure that functions as an ear.

Aurantiacus: orange colored; a mixture of yellow and red [chrome orange].

Aurate: with ears or ear-like expansions: also = auratus.

Auratus: golden yellow [pale cadmium yellow].

Aurelia: = chrysalis or pupa; specifically of butterflies.

Aurelian: a lepidopterist.

Aureolate: with a diffused colored ring.

Aureole: a ring of color which is usually diffuse outwardly.

Aureous -eus: gold-colored.

Aurichalceous: brassy yellow.

Auricle -cula: an appendage resembling a little ear; in Odonata the tumescent area at the sides of the second abdominal segment: in Andrenidae, a short membranous process placed laterally on the ligula.

Auricular: applied to the space or cavity surrounding the dorsal vessel.

Auriculate: with an ear-like appendage or, in antennae, with the basal joint distended into a concave, plate-like ear which envelops the rest of the structures.

Auriculo-ventricular: the outer valves of the heart between the auricular space and the chamber.

Auriculo-ventricular openings: are the lateral openings into the heart by means of which the blood is admitted into it.

Auritus: with two ear-like spots or appendages.

Auroral spot: applied to the bright orange colored spot at the apical area of Anthocharis.

Auroreous -eus: red, like the aurora borealis [crimson lake].

Austral: is that faunal region which covers the whole of the United States and Mexico except the boreal mountains and tropical lowlands: divided into transition, upper, lower and gulf strip: see boreal and tropical.

Austroriparian faunal area: that part of lower austral zone covering the greater part of the South Atlantic and Gulf States. Begins near mouth of Chesapeake Bay, covers half or more of Virginia, North and South Carolina, Georgia, Florida, Alabama, all of Mississippi and Louisiana, east Texas, nearly all of Indian Territory, more than half of Arkansas and parts of Oklahoma, s. e. Kansas, so. Missouri, so. Illinois, s. w. corner of Indiana and bottom lands of Kentucky and Tennessee.

Autotype: any specimen identified by the describer as an illustration of his species and compared with the type or co-type.

Auxiliary: additional, or supplementing.

Auxiliary vein: in Diptera (Will.), = subcosta (Comst.).

Axillae: two small, subtriangular sclerites at the lateral basal angles of the meso-scutellum in Proctytripidae.

Axillary: placed in the crotch or angle of origin of two bodies; arising from the angle of ramification.

Axillary area: see anal area.

Axillary calli: see calli axillary.

Axillary cell: in Diptera (Will.), = 2d anal (Comst.).

Axillary excision: = a. incision, q.v.

Axillary incision: Diptera; an incision on inner margin of wing, near base, which separates the alula from the main part.

Axillary lobe: the sclerite covering the base of the wing in Diptera; see also alula and posterior lobe.

Axillary vein: one or two longitudinal veins toward the inner margin from the anal vein (Ephemeridae); a group of several (10-20) radiate veins that occupy the anal field in Orthoptera.

Axis: a small process at base of elytron, upon which it turns.

Azure -eus: clear sky-blue [cobalt blue].

Azygos: unpaired; a structure without a fellow; sometimes applied to an unpaired oviduct specifically the enlarged portion of the vagina at the junction of the oviducts and thus = uterus.

B

Baccate -us: berry-like: applied to bladder-like ovaries from the surface of which the short ovarian tubes arise.

Back: the dorsum or upper surface.

Baculiform: rod or staff-like.

Badius: liver-brown; clearer and lighter than castaneus [dragon's blood].

Baenomere: a leg-bearing (thoracic) segment.

Baenopoda: the thoracic legs.

Baenosome: the thorax.

Balancers: see halteres.

Bald: without hair or other surface vestiture: see bare.

Band: a transverse marking broader than a line.

Bar: a short, straight band of equal width.

Barb: a spine armed with teeth pointing backward.

Barbate: furnished with barbs; hair with spines or spurs directed backward.

Barbated: bearded; in antennae with tufts or fascicles of hair or short bristles on each side of each joint; = brush-like: on the abdomen, with flat tufts at the sides or tip.

Barbule: a small barb, beard or filiform appendage.

Bare: without clothing of any kind: see bald.

Basad: in the direction of or toward the base.

Basal: at or pertaining to the base or point of attachment to or nearest the main body.

Basal area: in wings: that space nearest the point where they are attached to the body: on the metanotum of Hymenoptera, the anterior of the three median cells or areas = 1st median area.

Basal cell: Diptera; st (Will.), = radial 2 (Comst.); 2d (Will.), = media (Comst.); Trichoptera; one, two or three cells enclosed by the branches that form-the post-costal or anal vein: Odonata; an elongate cell between radius and cubitus, just before the arculus.

Basalis: the principal mandibular sclerite, when sclerites are distinguishable, to which all other parts are jointed; corresponds to the stipes in the maxilla.

Basal line: in many Lepidoptera; a transverse line extending half way across the primaries very close to base.

Basal lobe: of culicid genitalia, see claspette.

Basal post-costal vein: in Agrioninae, one of the cubito-anal cross- veins.

Basal segment of clasp: see side piece.

Basal space: that area on the primaries of certain Lepidoptera, between the base and t. a. line (q.v.).

Basal streak: in Noctuid moths, extends from base, through the submedian interspace to the t. a. line.

Basal transverse carina: on the metanotum of Hymenoptera, crosses before middle and separates the anterior from the median areas.

Base: that part of any appendage that is nearest the body: on the thorax that portion nearest the abdomen; on the abdomen that portion nearest the thorax.

Basement membrane: that thin layer of tissue upon which the epithelium rests.

Basilar: of or pertaining to the base.

Basilar cross-vein: Odonata; crosses the basilar space.

Basilar membrane: a thin membrane separating the cones and rods from the optic tract.

Basilar space: Odonata; that area at base of wings, between media and cubitus.

Basi-proboscis: basal third of the flexed proboscis of muscid flies.

Batesian mimicry: see mimicry.

Bathmis: see pterostigma.

Bave: the fluid silk as it is spun by caterpillars.

Beak: any notable prolongation of the front of the head: the snout in Rhynchophora: specifically, the jointed structure covering the lancets in the hemipterous mouth.

Bearded: fringed with hair: see barbated.

Belly: venter; under side of abdomen.

Belonoid: needle-like.

Bi: prefix, means two.

Bi-alar: two-winged; applied to Diptera.

Biarcuate: twice curved.

Biareolate: with two cells or areoles: see bilocular.

Bicaudate: having two tails or anal processes.

Bicolored: with two colors that contrast to some extent.

Bicornute: with two horns or cephalic processes.

Bicuspidate: ending in two points or cusps.

Bidactylate: with two fingers or finger-like processes.

Bidentate: two-toothed.

Biemarginate: twice emarginate; with two excisions.

Bifarious: pointing in opposite directions.

Bifasciate: with two bands or fascia.

Bifid: divided into two parts; split; applied in Coleoptera to tarsal claws which are divided so that the claws lie side by side: see biparted.

Biflabellate: antennae with fan-like process on two sides.

Bifurcate: divided, not over half its length, into two dull points; forked.

Bifurcation: a forking or division into two: the point at which a forking occurs.

Biguttate: with two drop-like spots.

Bijugum: in two pairs.

Bilamellar: divided into two lamina or plates.

Bilateral -eriter: with two equal or symmetrical sides.

Biliary vessels: see malpighian tubules.

Bilineate -us: with two lines.

Bilobate -ed: divided into two lobes.

Binocular: having two cells or compartments: see biareolate.

Bimaculate: with two spots or maculae.

Binate: in pairs: consisting of a single pair.

Binotate: with two rounded spots.

Binus: paired: doubled.

Biogenesis: the production of life from antecedent life.

Biomorphotica: those neuropterous insects in which the pupa is active.

Bionomics: the habits, breeding and adaptations of living forms.

Biophore: an ultimate constituent of germ plasm or hereditary substance.

Bioplasm: formative living matter.

Biparted: profoundly divided into two parts: see bifid.

Bipectinate: antennae having comb-like teeth or processes on Beach side of each joint.

Bipupillate: an ocellate spot with two pupils, of the same or different in color.

Biradiate: consisting of, or with two rays or spokes.

Biramose -ous: having two branches or doubled appendages.

Biseriately: arranged in double rows or series.

Biserrate: doubly saw-toothed; with a saw tooth on each side of each antennal joint.

Bisetose -ous: with two bristle-like or setaceous appendages.

Bisinuate: a margin or line with two sinuations or incisions.

Bituberculate: with two distinct tubercles.

Biuncinnate: with two hooks.

Bivalve -ed: applied to mouth parts consisting of two parts or valves united to form a tube.

Bivittate: with two longitudinal stripes or vittae.

Blade: of maxilla, see lacinia.

Blastem: a nucleated protoplasmic layer preceding the blastoderm.

Blastoderm: the germinal membrane from which the organs of the embryo are formed.

Blastodermic cells: are those forming the blastoderm.

Blastogenic: relating to or inherent in the germ or blast.

Blastophore: the primitive mouth of the embryo.

Blind: without eyes: applied also to an ocellate spot without a pupil.

Bloom: a fine violet dusting similar to that on plums.

Blotch: a large irregular spot or mark: large whitish membrane between abdomen and thorax in certain saw-flies.

Blunt: not sharp; obtuse at the edge or tip.

Body: the trunk: usually applied to the thorax only; rarely to the abdomen alone; sometimes to thorax and abdomen combined.

Bombifrons: front of head with a blister-like protuberance.

Bombous: blister-like; spherically enlarged or dilated.

Bombycinous: a very pale yellow like fresh spun silk.

Boreal: from or belonging to the north: is that faunal region that extends from the polar sea southward to near the northern boundary of the United States and farther south occupies a narrow strip along the Pacific Coast and the higher parts of the Sierra-Cascade, Rocky and Alleghany Mountain ranges; divided into Arctic, Hudsonian and Canadian: see austral and tropical.

Borer: applied to an insect or larva that burrows or makes channels in woody or other vegetable tissue.

Botryoidal: clustered like a bunch of grapes.

Bouclier: the pronotum, q.v.

Bouton: a button; the terminal lappet-like process at the tip of the ligula in bees: = spoon.

Brachelytra: with abbreviated wing covers or elytra.

Brachia: the arms: has been applied to raptorial fore-legs.

Brachial: relating to an arm; arm-like.

Brachial cells: Hymenoptera; 1st (Nort.), = costal and subcostal (Comst.) 2d (Nort.), = medial (Comst.); 3d (Nort.), = cubital (Comst.); 4th (Nort.), = 2d anal (Comst.).

Brachial veins: of primaries in Hymenoptera, originate at base, run parallel to inner edge toward anal angle; often connected with the cubital cellules by means of recurrent venules.

Brachium: the fore tibia.

Brachycerous: Diptera; with short, 3-jointed antennae.

Brachypterous: with short or abbreviated wings.

Brachyostomata: brachycerous Diptera with short proboscis.

Brain: that ganglion of the nervous system which lies in the head above the oesophagus; formed of the first three primitive ganglia: see supra-oesophageal.

Branchiae: air tubes or gill-like processes of aquatic larva;.

Branchial: relating to the gills or branchiae.

Branchiate: supplied with gills or bronchia.

Brassy: yellow, with the lustre of metallic brass.

Breast: the under surface of thorax or sternum.

Breast-bone: in Cecidomyid larvae; a horny, more or less elongate process of the under side behind the mouth opening, supposed to represent the labium = anchor process.

Breathing pores: see spiracle.

Brevis: short.

Brides: Homoptera; two pieces on the face, one each side of clypeus and lower part of front.

Bridge: Odonata; a secondary longitudinal vein connecting the radial sector (Comst.) with Mi + 2, apparently forming a continuous part of the radial sector; it is the proximal portion of the subnodal sector of de Selys and Hagen.

Bridge cross veins: Odonata; those cross veins, one or more in number, extending between M1 + 2 and the bridge (in de Selys between principal and subnodal sectors) proximal to the oblique vein.

Brin: the fluid silk thread from each salivary gland.

Bristle: a stiff hair, usually short and blunt.

Broken: interrupted in continuity; as a line or band.

Bronze: the color of old brass.

Brood: all the specimens that hatch at about one time, from eggs laid by one series of parents and which normally mature at about the same time.

Brunneus: a pure reddish dark brown [indian red].

Brush-like: antennae with the joints laterally produced and tufted with short hair or bristles: see barbated.

Buccal: relating to the mouth cavity; rarely to the cheeks.

Buccal appendages: the mouth parts excluding the labrum: see trophi. Buccal cavity: the mouth: = oral cavity.

Buccal fissure: the mouth slit or opening: the opening on each side of the mentum.

Buccate: blown up, distended; especially the cheeks.

Bucculae: little cheeks or distended areas.

Budding: applied to that form of agamic reproduction found in plant lice.

Bulla: a blister or blister-like structure: the shield-like sclerite that closes the opening to the trachea in lamellicorn larvae: in Ephemerida a part of the costal area of the fore wing toward the tip, which is slightly swollen forward and furnished with more cross veins than elsewhere; practically the stigma, q.v.

Bullate: blistered.

Bullule: a small blister.

Bursa: a pouch or sac: a wing pouch in male caddice flies and in connection with a stalked hair pencil.

Bursa copulatrix: the copulatory pouch of the female in some orders; a modification of the vagina.

C

Caducous-us: deciduous; easily detached or shed.

Caecal tubes or pouches: sac, or blind tube-like structures surrounding the chylific ventricle at its junction with the crop, and secreting a digestive ferment.

Caecum: a blind sac or tube-like structure serving as one of the caecal tubes or pouches: see coecum.

Caelate: a surface with plane elevations of varying forms.

Caeruleus -eous: light sky-blue [between lavender and cobalt blue] = coeruleus.

Caerulescent: with a tinge of sky-blue.

Caesius -eous: a pale dull blue-gray [blue-gray].

Caespiticolous: frequenting or living in grassy pastures or lawns.

Calathiform: shaped like a deep bowl.

Calcar -ium; pl. ia: a movable spur or spine-like process: specifically the spines at the apex of a tibia.

Calcarate -us: with a movable spur or spine-like process.

Caliciform: shaped like a cup or calyx.

Calipers: the anal forceps in Dermaptera.

Calli axillary: Odonata; thickenings at the bases of the wings; distinguished as anterior at the base of the costa, and posterior at the base of radius + medius and cubitus: = axillary calli.

Callosity: a thick swollen lump, harder than its surroundings: = callous: also a rather flattened elevation not necessarily harder than the surrounding tissue.

Callous: see callosity.

Callus: a small callosity.

Caltrops spines: the branched and otherwise specialized irritating spines in Limacodid larvae.

Calva: a skull-cap: = epicranium, q.v.

Calx: the distal end of the tibia; the curving basal portion of the first tarsal joint.

Calyculate: applied to antennae, whose cup-shaped joints are so arranged as to fit one into the other.

Calypter: Diptera; the alula or squama when it covers the haltere.

Calyptra: a hood or cap; see alula.

Calyptrate: those flies that have aluke or membranous scales above the halteres.

Calyx: the cap or crown of the mushroom bodies of the procerebrum: see also egg-calyx.

Campanulate: bell-shaped: more or less ventricose at the base and a little recurved at the margin.

Campestral: applied to species inhabiting open fields.

Campodeiform: applied to larval forms which, in their early stages at least, resemble Campodea: = leptitorm.

Canadian zone: is that part of the boreal region comprising the southern part of the great transcontinental coniferous forests of Canada, the northern parts of Maine, New Hampshire and Michigan, and a strip along the Pacific Coast reaching south to Cape Mendocino and the greater part of the high mountains of the United States and Mexico. In the east covers Green. Adirondack and Catskill Mountains and the higher mountains of Pennsylvania, West Virginia, Virginia, western North Carolina and eastern Tennessee. In the Rockies extends continuously from British Columbia to western Wyoming and in the Cascades from British Columbia to southern Oregon with a narrow interruption along the Columbia River.

Canaliculate: channelled; longitudinally grooved, with a deeper concave line in the middle.

Cancellate: cross-barred: latticed: with longitudinal lines decussate by transverse lines.

Canescent: hoary, with more white than gray.

Canine teeth: applied to the sharp and conical teeth of mandibles in predatory species: = dentes caninae.

Cantharidin: the substance that gives the meloid beetles their blistering power composition, $C_{10}H_{12}O_4$ (von Furth).

Canthus: the chitinous process more or less completely dividing the eyes of some insects into an upper and lower half.

Canus: see canescent.

Capillaceous: capilla or hair-like.

Capillaris: a very slender, hair-like tube.

Capillary: long and slender like a hair: antennae in which the joints are long, slender and loosely articulated.

Capillate -us: clothed with long slender hair; = coryphatus.

Capillii: hairs of the head that form a cap as in certain Trichoptera and Tineid Lepidoptera.

Capillitium: the hood-like collar in some Noctuid moths, e.g. Cucullia: see cucullus.

Capitate: with a head: that type of clavate antenna in which the club is abruptly enlarged at tip and forms a spherical mass.

Capitulum: a small head: the enlarged tip of an antenna: the little knob at tip of halteres in Diptera: the labella or lapping tip of the mouth of certain flies.

Capricorn beetle: a Cerambycid or long horned beetle.

Caprification: is that method or process through which the Smyrna figs are fertilized by Blastophaga throughthe medium of wild, inedible or "caprifigs."

Capsular: in the form of a capsule or little cup-like container.

Caput: the head with all its appendages.

Capylus: a hump on the Tupper side of the segments of many larva.

Carabidoid: applied to the second stage of a meloid larva, when it resembles that of a Carabid.

Carbonarius: coal black.

Cardia: the gizzard; q.v.: also applied to the heart.

Cardiac: belonging or relating to the heart.

Cardiac valvule: see oesophageal valve.

Cardinal cell: Odonata; see triangle.

Cardioblasts: a string or row of cells in the embryo giving rise to the heart or dorsal vessel.

Cardio-coelom: that part of the coelom that forms the pericardium.

Cardio-coelomic: applied to the venous openings from the heart to the body cavity.

Cardo, pl. Cardines: the hinge or basal sclerite of the maxilla by means of which it is jointed to the head.

Carina -ae: an elevated ridge or keel, not necessarily high or acute. Carinate: a surface having carinae.

Carinula -ae: a little carina or keel-like ridge; specifically, the longitudinal elevation on the middle of snout in Rhynchophora.

Carinulate: a surface with small and rather numerous carinae.

Cariose -ous: corroded; appearing as if worm-eaten.

Carminate -ed: mixed or tinged with carmine.

Carneous -eus: flesh-colored [salmon with a little carmine].

Carnivorous: a feeder upon flesh food.

Cariose -us: of a soft, fleshy substance.

Carolinian faunal area: that area of the upper austral zone comprising the larger part of the Middle States (except the mountains), s. e. So. Dakota, east. Nebraska, Kansas and part of Oklahoma; nearly all of Iowa, Missouri, Illinois, Indiana, Ohio, Maryland and Delaware; more than half of West Virginia, Kentucky, Tennessee and New Jersey and large areas in Alabama, Georgia, the Carolinas, Virginia, Pennsylvania, New York, Michigan and South Ontario: extends along Atlantic Coast from near mouth of Chesapeake Bay to Southern Connecticut and sends narrow arms up the valleys of the Hudson and Connecticut. A narrow arm follows the east shore of Lake Michigan to Grand Traverse Bay.

Carpus: the pterostigma of Odonata: the extremity of the radius and cubitus of the primaries: that point in the wings at which they are tratsversely folded.

Cartilaginous: of the consistency of cartilage or gristle.

Caruncle: a soft, naked, fleshy excrescence or protuberance.

Caryophylleous: nut or clove brown [Indian red].

Castaneous: chestnut brown; bright red-brown [dragon's blood with a slight admixture of vermilion].

Castes: the various forms or kinds of matured individuals among social insects as workers, soldiers, queens, etc.

Cataphracted: invested with a hard callous skin, or with scales closely united. Catch: in Collembola, = tenaculum, q.v.

Catenate: with longitudinal connected elevations like links in a chain.

Catenulate: like catenate; but the links are smaller.

Caterpillar: the term applied to the larvae of Lepidoptera.

Catervatum: by heaps.

Caudal: the tail: any process resembling a tail: the pointed end of the abdomen in plant lice: any extension of the anal segment or appendage terminating the abdomen.

Caudad: toward the posterior end of the body, along the median line.

Caudal: pertaining to the posterior or anal extremity.

Caudal setae: long, thread-like processes at the end of the abdomen in many europterous and some other insects; = anal filaments.

Caudate: with tail-like extensions or processes.

Caudo-cephalic: in a line from the head to the tail.

Caudo-dorsad: directed upward and toward the tail.

Caudula -ae: a little tail.

Caul: the fatty mass of larvae from which the organs of the future adult were supposed to develop: = epiploon.

Cauliculus: the larger of the two stalks supporting the calyx of the mushroom body.

Caulis: the funicle of antenna: the corneous basal part of jaws.

Cavate: hollowed out; cave-like.

Cavernicolous: cave-inhabiting.

Cavernous: divided into small spaces or little caverns.

Cavity -as: a hollow space or opening.

Cecidium: a gall.

Cell: any space between or bounded by veins: in the Comstock system the cells derive their names from the vein forming the Tupper margin: e.g. all just below the radius are radial cells; and they are numbered from the base outward, as radial 1, 2, etc.: the living unit; protoplasm differentiated into cytoplasm and nucleus, from which units all but the lowest plants and animals are developed by division and consequent increase into a multicellular condition: a compartment or division of a nest or honey-comb.

Cellule: a portion of a wing included between veins; usually applied to a small area completely inclosed, rarely to interspaces where no closed area is formed.

Cenchrus -rib: minute, often white marks, or membranous spaces on the metanotum of some Hymenoptera.

Cenogonous: producing young at one time oviparously, at another viviparously as in plant-lice.

Centimeter: abb. Cm.: = .01 meter = .394 inch; 2.54 Cm. = one inch.

Centrad: toward the centre or interior.

Central foveola: see median foveola.

Centrolecithal: applied to eggs in which the food yolk is central.

Centrosome: a spherical body that appears outside the nucleus of a cell.

Cephalad: toward the head, along the central line of the body.

Cephalic: belonging or attached to the head; directed toward the head.

Cephalic bristles: Diptera; specialized bristles occurring on the head.

Cephalic foramen: the posterior or occipital foramen of head through which the dorsal vessel, oesophagus, salivary ducts and ventral nerve cords pass from head to prothorax.

Cephalization: concentration toward the head.

Cephalomere: one of the head segments of an arthropod.

Cephalophragm: a v-shaped partition which divides the head of some Orthoptera, into an anterior and posterior chamber.

Cephalon: the head.

Cephalosome: the head as one of the three regions.

Cephalotheca: the head covering in the pupal stage.

Cephalotheca: the united head and thorax of arachnids and crustacea: that portion of an obtect pupa covering head and thorax: the anterior segments of larva that have no obviously separated head.

Cerago: bee-bread.

Ceratheca or Ceratotheca: that portion of the pupal shell that envelops the antenna.

Cerci: two lateral anal appendages; usually short, jointed, antenna- like, developed from the eleventh abdominal segment of the embryo; sometimes unjointed and specialized into forceps or other processes.

Cercopoda: jointed foot-like appendages of the last abdominal segment; also applied like cerci.

Cercus: see cerci.

Cerebellum: has been applied to the sub-esophageal ganglion.

Cerebrum: the supra-oesophageal ganglion.

Cernuous: bent: with the apex bent downward.

Cervical: relating or belonging to the neck.

Cervical foramen: in coleopterous larvae - occipital foramen.

Cervical sclerites: small ebitinous plates on the membrane between head and thorax: see jugular sclerites.

Cervical shield: the ebitinous plate on the prothorax of caterpillars just behind the head: = prothorax shield.

Cerviculate: with a long neck or neck-like portion.

Ceryinus: reddish, deer-gray [pale cadmium yellow and Indian red].

Cervix: the upper part of the neck; = crag: in Diptera; that part of the occiput lying over the junction of the head, i.e. between the vertex and neck.

Cespitose: matted together.

Chaetophorous: applied to bristle-bearing flies.

Chaetotaxy: the science dealing with the arrangement and nomenclature of the bristles on the body of insects.

Chagrined: see shagreened.

Chalastrogastra: the saw-flies: a group of Hymenoptera.

Chalceous: brassy in color or appearance.

Chalybeate: steely in appearance.

Chalybeous: metallic steel blue.

Channelled: a surface, with deep grooves or channels.

Chaperon: =clypeus or clypeus anterior.

Chaplet: a little crown; a circle of hooks or other small processes terminating a member or appendage.

Character: a quality of form, color or structure.

Cheek: see gena.

Chela: the terminal portion of a limb bearing a lateral movable claw like that of a crab; specifically applied to the feet in some Parasitica in which the opposable claw forms a clasping structure.

Chelate: bearing a cheat or claw; applied when claws are capable of being drawn down or back upon the last tarsal joint.

Chiasma: an X-like crossing of nerve fibers.

Chirotype: a specimen upon which a manuscript name is based.

Chitin: the material forming the hard parts of the insect body; it is a secretion (or a metamorphosis?) of the epidermis, differing from horn by its insolubility in boiling liquor potassae: = elytra, entomolin.

Chitinogenous: applied to that layer of epidermal cells which secretes the chitin.

Chitinization: the process of depositing or filling with chitin.

Chitinized: filled in with or hardened by chitin.

Chitinous: composed of chitine {*Scanner's comment: sic*} or like it in texture: as a color term is amber yellow.

Chlorophane: an oily, greenish yellow pigment found in insects.

Chlorophyll: the green coloring matter of plants; one of the substances found in the blood of insects.

Chordotonal: responsive to vibrations; applied to the ear-like structures in Orthoptera,.

Chorion: the shell or covering membrane of an insect egg.

Chromatin: the minute granules that make up the chromoplasm of a cell nucleus.

Chromosome: one of the segments into which the chromoplasmic filaments of a cell nucleus breaks up just before indirect division.

Chrysalis or -id: applied specifically to the intermedial stage between larva and adult in butterflies: see pupa.

Chrysargyrus: silvery gilt.

Chyle: the food-mass after it has passed through the guard and is mixed with the secretions of the salivary glands and caecal structures, ready to be assimilated.

Chylific ventricle: the true stomach in which the chyle is prepared and digestion begins.

Cibarian: referring to the mouth parts.

Cicatricose: a surface having scars with elevated margins like those of smallpox.

Cicatrix: a scar: an elevated, rigid spot.

Cilia: fringes; series of moderate or thin hair arranged in tufts or single lines; thin scattered hair on a surface or margin.

Ciliate: fringed: set with even, parallel hairs or soft bristles.

Cilium, pl. Cilia: q.v.

Cimicine: an oily fluid of disagreeable odor secreted by certain Heteroptera and used as a means of defense.

Cimier: the head crest in Pierid chrysalids.

Cinetus: with a colored band:= cingulatus.

Cinereous: ash-colored; gray tinged with blackish [ultra ash gray]. Cinerescent: ashen in color or appearance.

Cingula -um: a colored band or bands.

Circulate -us: having a cingulum or collar: see also cinetus.

Cinnabarine: [vermilion red].

Cinnamomeous: cinnamon brown [burnt sienna].

Cinema: see Thysanura, of which this forms a group including the bristle-tails, and for which it has been used as an equivalent.

Circinal: spirally rolled like a watch-spring or a butterfly tongue.

Circiter: about, or round-about.

Circular: round like a circle.

Circumgenital glands: small circular glands with an excretory orifice at tip, disposed in groups about the genital orifice in Diaspinae.

Circumoesophageal commissures: those cords or nerve fibres connecting the suboesophageal ganglion with the main trunk of nervous system.

Circumsepted: with a vein all around the wing.

Citrate: antennae with very long, curled lateral branches which may or may not be ciliated; see plumose.

Cirrose -us: with somewhat dense curled hair.

Cirrus: a curled lock of hair placed on a thin stalk.

Citrine -us: lemon yellow [chrome yellow].

Cladocerous: with branched horns or antennae.

Clasper: a chitinized process, free or attached to the inner sides of harpes, valves or other lateral pieces, serving to hold the female parts during copulation: = the harpers of some authors.

Claspette: in genitalia of male culicids, the inner basal lobe of side piece; q.v.

Clasp-filament: in male genitalia of culicids the articulated appendage or terminal segment of side-piece or clasp; sometimes bears an articulated point or apex and then = articulated apex.

Class: a division of the animal kingdom lower than a sub-kingdom and higher than an order: e.g. the "Class Insecta."

Classification: is the systematic arrangement of insects (or other animals or plants) in series showing their relation or agreement in structure, life habits or other characters forming the basis of the "classification."

Clathrate: latticed or lattice-like in appearance.

Claustrum: the structure uniting the wings in flight, whether by hooks, by a thickening of the margin, or by a jugum.

Clava: a club; the enlarged apical joints of a clubbed antenna: = clavola.

Claval suture: Hemiptera; at the base of hemelytra, separating the clavus.

Clavate: clubbed: thickening gradually toward the tip.

Clavate hairs: in Collembola, = tenent hairs.

Clavicornia: that series of beetles having the antennae more or less distinctly enlarged or clubbed at tip.

Clavicular lobe: Homoptera; that portion of hind wing behind anal veins.

Claviform: club-like in form; specifically, in Noctuid moths an elongate spot or mark extending from the t. a. line through the submedian interspace, toward and sometimes to the t.p. line.

Clavola: see clava.

Clavus: the club of an antenna lava and clavola: in Heteroptera, the oblong sclerite at the base of the inferior margin of the hemelytra: the knob at the end of the stigmal or radial veins in certain Hymenoptera.

Claws: the claw or hook-like structures at the end of the foot or tarsus.

Cleavage: see segmentation of egg.

Cleft: split: partly divided, longitudinally: in Coleopteran applied to claws so divided that the parts lie one above the other.

Clintheriform: shaped like a plate. Cloaca: see rectum.

Clubbed: see clavate.

Clypeal suture: marks the division between clypeus and epicranium.

Clypeate: shield-like in form.

Clypeate constriction: applied when a surface is drawn in from the sides so as to produce a shield or saddle-like form.

Clypeo-frontal suture: = clypeal suture.

Clypeus: that portion of the head before or below the front, to which the labrum is attached anteriorly; in Diptera often visible below the margin of the mouth in front, as a more or less visor-shaped piece:= epistoma.

Clypeus-anterior: see ante-clypeus.

Clypeus posterior: see post-clypeus.

Coactus: condensed; of a short stout form.

Coadapted: formed so as to work together to one end; as the mandible and maxilla in Chrysopids, etc.

Coadunate: joined together at base; two or more joined together; said of elytra when permanently united at the suture.

Coagulate: to congeal; to change from a fluid to a jelly

Coagulum: a clotted mass, as of blood.

Coalescent: united or grown together.

Coarctate: contracted: compacted: applied to that form of pupa in which all the members of the future adult are concealed by a thickened, usually cylindric case or covering, which is often the hardened skin of the larva: beginning with a narrow base, then dilated and thickened.

Cocardes: retractile vesicular bodies on each side of the thorax in certain Malachidae.

Coccineous: cochineal red; dark red [carmine].

Cochleiformis: formed like a snail shell.

Cochleate: spirally twisted like a screw or a univalve shell.

Cocoon: a covering, composed partly or wholly of silk or other viscid fibre, spun or constructed by many larvae as a protection to the pupa.

Cocoon-breaker: structures or processes of the pupa, often on the head, by means of which it works its way out of the cocoon.

Coecal: ending blindly, or in a closed tube or pouch.

Coecum: a blind sac or tube: applied to a series of appendages opening into the alimentary canal at the junction of the gizzard and chylific ventricle: see caecum; the two are used interchangeably.

Coeloblast: the endoderm in the narrower sense.

Coelom: the body cavity.

Coelomic cavity: the space between the viscera and the body wall.

Coelom-sac: the cavity containing the viscera: in embryology one of a pair of closed sacs, arising in the mesoderm of each segment of the embryo and giving rise to more or less of the coelom of the adult.

Coenogonous: oviparous at one season of the year, ovoviviparous at another, as in Aphididae.

Coeruleus -eous: sky-blue: see caeruleus.

Coincident: when two wing veins run together or lie, one in continuation of the other so as to appear like one.

Coleoptera: sheath-winged: an order with the primaries coriaceous, used as a cover only, meeting in a straight line dorsally; mouth mandibulate; pro-thorax free; transformation complete: the beetles: the term has also been applied to the two elytra together.

Collar: in general any structure between the head and thorax: specifically, in Hymenoptera, the neck; in Diptera, may mean the neck, the sclerites attached to the thorax, the thorax itself, or its processes (ante furca): in Coleoptera, is the narrowed thorax; in Lepidoptera, applied to the sclerites attached to the thorax and which shield the neck.

Collembola: an ordinal term applied to species which are apterous; have no metamorphoses; have variably developed abdominal saltatorial appendages and a peculiar ventral tube at base: the spring-tails.

Colleterial gland: see Colleterium.

Colleterium: a glandular structure accessory to the oviduct, secreting the viscid material used in cementing the eggs together.

Collophore: the sucker-like organ extended from the underside of the abdomen in Collembola.

Collum: the neck or collar: the slender connection between head and thorax in Hymenoptera and Diptera; in Coleoptera, the posterior, narrow part of the head or even the thorax: loosely used.

Colon: the large intestine; that usually enlarged portion of the alimentary canal before the rectum.

Columella: a little rod, pillar or central axis.

Columnar: cylindric, but tapering toward one end.

Comate -us: only the upper part of head, or vertex, covered with hair. Commensal: one who eats at another's table: applied to species that feed on the surplus supply of another, without destroying the owner of the supply.

Commensalism: applied to this manner of living and eating together.

Comminute: to grind up fine: to reduce to minute particles.

Commissure: the nerves connecting two ganglia: the point of meeting or union of two bodies: a bridge connecting two bodies or structures; e.g. tracheal tubes.

Common: of frequent occurrence: occurring on two adjacent parts: a band or fascia is common when it crosses both primaries and secondaries.

Communal: applied to life or dwelling in colonies like ants and bees.

Comose: ending in a tuft or brush.

Complanate: compressed: flattened above and below: = deplanate.

Complemental: applied to sexed forms in the Termitidae, capable of reproduction, but which do not reach the winged stage; the females are less fertile than the forms that become winged and several may be used in one nest to replace a lost queen or mature female.

Complicant: when one elytron extends over the other and partially covers it.

Complicate: longitudinally laid in folds: intricate as opposed to simple.

Component: one part of a combined whole.

Compound: made up of many similar or dissimilar parts.

Compressed: flattened laterally.

Concatenate: linked together in a chain-like series.

Concave: hollowed out; the interior of a sphere as opposed to the outer or convex surface: concave veins are those that occupy the bottoms of troughs or grooves on the upper surface of a wing; see convex veins.

Concavo-convex: hollowed out or concave on one surface, rounded or convex on the other; like a small segment of a hollow sphere.

Concentrated: gathered together at one point; intensified or strengthened by evaporation.

Conchate: applied to the shell-like inflation of the auricle in the cephalic tibia of Orthoptera.

Concinne: neat; fine.

Concolorous: of the same general color.

Concretion: a massing together of parts or particles.

Concurrent: applied to a vein which arises separately, runs into another and does-not again separate.

Conduplicate: doubled or folded together.

Condyle: a process which articulates the base of the mandible to the head: in general any process by means of which an appendage is articulated into a pan or cavity.

Confertim: closely clustered or crowded.

Conflect: crowded; clustered; opposed to sparse.

Confluent: running together; as of two macula when united in one outline.

Confused: a marking with indefinite outlines: a running together as of lines and spots without definite pattern.

Congener: a species belonging to the same genus.

Congeneric: applied to a species agreeing in all characters of generic value with others compared with it.

Congested: heaped together; crowded: distended.

Conglobate: gathered together in a ball or sphere.

Conglobate gland: a glandular appendage of male sexual organs in Orthoptera, opening upon one of the external structures.

Conglomerate: congregated; massed together.

Conic -al: cylindrical, with a flat base, tapering to a point.

Conico-acuminate: in the form of a long, pointed cone.

Coniferous: a surface which bears cone-like processes.

Conjugate: to bring together in pairs: consisting of a single pair. Conjugation: the union of pairs; usually applied to the merging of the male and female elements.

Conjunctiva: the membrane uniting the abdominal sclerites.

Conjunctivus: a mandibular sclerite between the molar and basalis.

Conjunctura: the articulation of a wing to the thorax.

Connate: united at base, or along the whole length.

Connexivum: the prominent abdominal margin of Het., at junction of dorsal and ventral plates: also used like pulmonarium, q.v.

Connivent: converging: approaching together: wings so folded in repose that they unite perfectly at their corresponding margins.

Consperse: irregularly dotted or sprinkled.

Conspicuous: striking: easily seen at a glance.

Conspurcatus: confusedly sprinkled with discolored or dark spots.

Constituent: a part or element of a whole.

Constricted: drawn in: narrowed medially and dilated toward the extremities.

Contiguous: so near together as to touch.

Contorted: twisted: obliquely incumbent upon each other.

Contour: the outline or periphery.

Contract -ed: to draw or drawn together: to reduce, or reduced in size by contraction.

Contractile: that which may be drawn together or contracted or which has the power of contracting.

Contrasting: appearing in sharp relief or contrast; as one color or marking against another.

Converging: approaching each other toward the tip.

Convergence: the approaching or drawing together at tips.

Convex: the outer curved surface of a segment of a sphere; opposed to concave: convex veins are those which occupy the summits of ridges on the upper surface of - wing; see concave veins.

Convolute: rolled or twisted spirally: also applied to wings when they are wrapped around the body.

Coprophagus: feeding on excrement or on decaying vegetable matter of an excrementitious character.

Copula, Copulation: the act of sexual union.

Copulate: to unite in sexual intercourse.

Copulation chamber: a chamber or cell excavated by certain Scolytid beetles in their burrows, in which copulation takes place: = rammel-kammer.

Coralline: a pale pinkish red [salmon].

Corbel: an ovate area at the distal end of the tibia in Coleoptera, surrounded by a fringe of minute bristles; when the articular cavity is on the side, above the tip, the corbel is closed; when the cavity is at the extreme tip, the corbel is open.

Corbicula -um: a concave, smooth space, edged by a fringe of hairs arising from the margins of the posterior tibiae in bees, forming the pollen basket its function is to hold the collected pollen in place.

Corbiculate: having corbicula.

Cordate: heart-shaped; triangular, with the corners of the base rounded: not necessarily emarginate at the middle of base.

Cordiform: = cordate.

Coriaceo-reticulate: with impressed reticulations giving a leather-like appearance.

Coriaceous: leather-like: thick, tough and somewhat rigid.

Coriarious: leather-like in sculpture or texture.

Corium: the elongate middle section of the hemelytra which extends from base to membrane below the embolium.

Cornea: the outer surface of the compound eye as a whole, and of each individual facet.

Corneal lenses: are the individual lens-like structures of which the cornea of the compound eye is composed.

Corneous: of a horny or chitinous substance; resembling horn in texture.

Cornicles: the honey tubes in plant-lice: = corniculus.

Corniculi: the little horny tips or pieces of the ovipositor in Orthoptera; see valves.

Corniculus -i: = cornicles; honey-tubes; q.v.

Corniform: like the horn of an ox: a long, mucronate or pointed process.

Cornute -us: having horns or horn-like processes.

Corona: a crown or crown-like processes.

Coronate: with a crown-like tip or termination.

Coronet: a small crown or corona.

Coronula: a circle or semicircle of spines at the apex of the tibia.

Corpus: the body as a whole.

Corpus adiposum: the mass of fat tissue often found in larvae.

Corpuscle: a small cell; usually applied to blood cells.

Correlate: to bring together into relation or correspondence.

Correlated: derived from the same ancestral form: said of two or more features or qualities which bear a direct or an inverse relation to each other, but without implying a relation of cause and effect.

Correlative: of a correlated nature; see correlated.

Corrode: to eat away gradually, as by rust or decay.

Corrodentia: an ordinal term meaning gnawers: net-veined or wingless: mandibulate, mouth formed for gnawing; transformation incomplete; thorax incompletely agglutinated: = Psocoptera: includes Termitidae, Psocidae and Mallophaga.

Corrugated: wrinkled; with alternate ridges and channels.

Corselet: the thorax in Coleoptera.

Cortical: relating to the cortex or outer skin.

Corticinus: bark-like in sculpture, texture or color [vandyke brown].

Corvinus: crow-black; deep, shining black with a greenish lustre.

Coryphatus: = capillatus.

Corysterium: an abdominal glandular structure in certain females, secreting a glutinous covering for the eggs.

Cosmopolitan: species that occur throughout most of the world.

Cosmotropical: species that occur throughout the tropics.

Costa: any elevated ridge that is rounded at its crest: the thickened anterior margin of any wing, but usually the primaries: in Comstock, the vein extending along the anterior margin of the wing from base to the point of junction with subcosta.

Costal area: the area behind costal vein; see also, costal field.

Costal cell: the area inclosed between the costal and subcostal veins: in the plural, Comstock, are all the cells anteriorly margined by the costa; in Hymenoptera (Norton), includes the 1st, 2d and subcostal; of Packard, the 3d costal = 2d radial 1, and radial 2: in Diptera (Will.), it is the 2d costal.

Costal field: Orthoptera; that region of the tegmina adjacent to the anterior margin or costa: = anterior field.

Costal fold: in the males of some Hesperidae, a membranous flap that may be opened to expose the androconia.

Costal margin: the anterior margin of a wing whether it is really costate or not.

Costal membrane: Hymenoptera; the surface of wing in front of costal vein.

Costal vein: Lepidoptera; runs close to and parallel with the costal margin, extending from base to the margin before the apex; always simple and often absent in the secondaries; is vein 12 of the numerical series on primaries; vein 8 on secondaries: = subcosta (Comst.).

Costate: ribbed; marked with elevated thickened lines.

Costula: Hymenoptera; a small ridge separating the externo-median meta-thoracic area into two parts.

Costulatus: less prominently ribbed than costate.

Cotyla: the articular pan; the cup or socket of a ball and socket joint.

Cotypes: are all the specimens before the describer when a species is named, no single one being selected as the type: the type in such case equals the sum of the cotypes: see paratype.

Coxa -ae: the basal segment of the leg, by means of which it is articulated to the body.

Coxal cavity: the opening or space in which the Coxa articulates; in Coleoptera the cavity is open when the epimera do not extend to the sternum; closed or entire when the epimera reach the sternum or join medially as in Rhynchophora; the cavities are separated when the prosternum extends between them, confluent when it does not: see acetabulum.

Coxal glands: eversible glandular structures at base of legs; well developed in some Thysanurans, modified variously in higher orders.

Coxal stylets: short, leg-like, jointed appendages on the underside of the abdominal segments in Thysanura.

Crag: the neck: = cervix.

Cranium: the head or skull except the neck; sometimes limited to the fixed parts above the clypeo-frontal suture.

Crassus: thick; tumid.

Crateriform: like a shallow funnel or deep bowl.

Creber: closely set.

Cremaster: a stout spine, process or hooked area at the hind end of pupae in Lepidoptera.

Crenate: scalloped, with rounded teeth.

Crenulate: with small scallops, evenly rounded and rather deeply curved.

Crepitation: a crackling sound or the production of such as by discharge of vapor or "bombarding": a cracking or creaking.

Crepuscular: active or flying at dusk.

Crescentiform: like a lunule or crescent.

Crescentric: lunulate.

Crest: a prominent, longitudinal carina on the upper surface of any part of the head or body.

Crested: see cristate.

Cretaceous: chalky white: the third, uppermost and latest of the three great divisions of the mesozoic or secondary rocks.

Cribrate: pierced with closely set, small holes.

Cribriform: with perforations like those of a sieve.

Crineous: dark-brown, with a slight admixture of yellow and gray.

Crinite -us: with tufts of long thin hair: see lanuginose.

Crispate -us: with a wrinkled or fluted margin.

Crista: a ridge or crest.

Cristate: with a prominent carina or crest on the upper surface::= crested.

Cristiform: in the form of a sharp ridge or crest.

Cristula: a small crest.

Cristulate: with little crescent-like ridges or crests.

Croceous: saffron yellow; yellow with an admixture of red [pale cadmium yellow].

Crocus: =croceous.

Crook: the hook or recurved tip of the antenna in Hesperidae.

Crop: the dilated portion of the alimentary canal behind the gullet which serves to receive and hold the food previous to its slower passage through the digestive tract: = ingluvies.

Crotchets: the curved spines or hooks on the prolegs of caterpillars and on the cremaster of pupae.

Crown: the top of head in Lepidoptera; also used as = coronet or corona.

Cruciate: shaped like a cross; applied to wings when the inner margins lie one over the other; or to incumbent wings that overlie only at the apex: in Diptera, applied to bristles when they cross in direction.

Cruciato-complicatus: folded crosswise: incumbent wings when the inner margins overlap; not well distinguished from cruciate.

Crura: the legs or, more specifically, the thighs.

Crura cerebri: two large cords that connect the supra- with the sub-oesophageal ganglion.

Crus: a leg or leg-like structure.

Crustaceous: hard, like the shell of a crab.

Crypto: hidden, concealed.

Cryptocerata: a division of Heteroptera with small antennae concealed in a groove under the bead: = adeloceratous: see gymnocerata.

Cryptogastra: with the venter or belly covered or concealed.

Cryptopentamera: feet 5-jointed, the 4th joint small and concealed.

Cryptotetramera: feet 4-jointed, one of them small and concealed.

Cryptothorax: a supposed thoracic ring between meso- and meta-thorax.

Crypts: minute secretory follicles or cavities: specifically, large gland- like structures between the epithelial cells in chylific ventricle.

Crystalline: transparent, like crystal.

Crystalline cone: a conical structure below the cornea, imbedded in pigment cells of the compound eye: also termed Crystalline lens.

Ctenidium: a comb-like structure occurring on any part of an insect.

Cubital: referring or belonging to the cubits.

Cubital cell: the wing area between the cubits and anal vein; in the plural, all the cells bounded anteriorly by the cubits or its branches (Comst.); in Diptera (Schiner), = radial 3 (Comst.), = 3d posterior cell (Loew); in Hymenoptera (Norton), = radial 3, 4 and 5 (Comst.).

Cubital forks: the branching or points of separation of the branches of the cubits.

Cubital nerve or vein: see cubits.

Cubitus: of Comstock, is the 5th in the series of longitudinal veins extending from base, and usually two branched before reaching outer margin: in Orthoptera; = the internomedian and ulnar: in Neuroptera, a main longitudinal vein next behind the medius and before the anal: the tibia of the anterior leg.

Cuckoo spit: liquid in the form of bubbles produced by members of the family Cercopidae and which often conceals the producer.

Cucullate: hooded; somewhat hood-shaped.

Cucullus: a hood: see capillitium.

Cuilleron: see alula.

Culicifuge: any preparation for driving away gnats or mosquitoes.

Culmen: the longitudinal carina of a caterpillar.

Cultellus: one of the blade-like lancets in piercing flies: = the mandibles of some authors.

Cultrate -iform: shaped like a pruning knife.

Cumulate: in groups or heaps.

Cumulus: a group or heap; as of cells in a developing ovum.

Cuneate, Cuneiform: wedge-shaped; elongate triangular.

Cuneus: Hymenoptera; the small triangular area at the end of the embolium of hemelytra: Odonata, the small triangle of the vertex between the compound eyes.

Cupreous: the metallic red of pure shining copper.

Cupules: the sucker-like processes covering the under surface of the tarsi in male Dytiscidce.

Cupuliform: cup-shaped: like a little cup: = cyathiform.

Cursoria: in Orthoptera, that series in which the legs are formed for running (roaches, etc.).

Cursorial: formed for running.

Curvate: curved.

Curvinervate: wings with the veins distinctly curved, like some Psocidae.

Cusp -is: a pointed process; sometimes at the margin of a wing.

Cuspidate: prickly pointed; ending in a sharp point; with an acuminated point ending in a bristle.

Custodite -us: guarded: a body in an envelope.

Cuticle: the outer skin or skin layer.

Cuticula: = cuticle: specifically applied to the outer or chitinized layer: see epidermis and hypodermis.

Cyaneous: pure dark blue: indigo blue [French blue].

Cyanescent: with a deep bluish tinge or shading.

Cyanogenic: applied to repugnatorial glands in myriapods and sometimes in insects.

Cyathiform: obconical and concave; cup-shaped: = cupuliform.

Cyatotheca: the cover of the thorax in the pupa.

Cycle: a round or circle, e.g. of development; a life cycle.

Cyclorrhapha: that section of Diptera in which the adult escapes from the hardened pupal case by pushing off a lid or covering: see orthorrhapha. Cyclorrhaphous: circular seamed.

Cydariform: globose, but truncated at two opposite sides.

Cylindrical: in the form of a cylinder or tube; round, elongate, of equal diameter throughout.

Cymbiform: boat-shaped: a concave disc with elevated margin; navicular.

Cytoplasm: the protoplasm of a cell exclusive of nucleus; the cell body.

D

Dactylus: a finger or toe: = digitus: a tarsal joint after the first one, when that is enlarged as in bees.

Dagger mark: a marking in the form of a Greek Psi Y .

Dart: a sting, or its central part.

Dash: a short disconnected streak or mark.

Dasygastres: bees with pollen-carrying structures on the abdomen.

Deaurate: of the color of gold; golden.

Deciduous: that which may be cast off or shed.

Declinate -us: a part somewhat bent, the apex downward.

Decumbent -ous: sloping gradually downward.

Decrepitans: crackling.

Decumbent: bending down at tip from an upright base.

Decurrent: closely attached to and running down another body.

Decurved: bowed downward.

Decussate: crossing at an angle: X-like: in cross pairs; or, when bristles alternately cross each other, as in some Diptera.

Deflected: bent downward: the wings, when the inner margins lap and the outer edges decline toward the sides.

Deflexed: abruptly bent downward.

Deformed: twisted or set in an unusual form: specifically, in Coleoptera applied to knotted or twisted antennae as in male Meloids.

Dehiscence: the splitting of the pupal integument in the emergence of the adult in Lepidoptera.

Dehiscent: open or standing open: separating toward the tip.

Dejectamenta: the excrement or excretion.

Delamination: the splitting or division into layers.

Deltoid: elongate triangular: resembling a Greek D with apex extended.

Demarcation: the bounding, laying out or limiting.

Dendritic: applied to the branched nerve cells in the mushroom bodies of the pro-cerebrum.

Dendroid: tree or shrub-like: branching like a tree or shrub.

Dendrophagus: feeding on woody tissues.

Dendrophilous: species that live in woody tissue, or on trees.

Dens: a tooth or tooth-like process.

Dense: thickly crowded together.

Dentate: toothed: with acute teeth, the sides of which are equal and the tip is above the middle of base.

Dentate-serrate: toothed, with the dentations themselves serrated on their edges.

Dentate-sinuate: toothed and indented.

Dentes: the teeth or pointed processes on the inner side of the mandible: the second or middle part of the furcula in Collembola, consisting of two parallel pieces from the distal end of the manubrium and bearing at their apices the crones.

Dentes caninae: see canine teeth.

Denticle: a small tooth.

Denticulated: set with little teeth or notches.

Dentiform: formed or appearing like a tooth.

Denudate: without covering; destitute of scales or hair.

Denude: to free from covering; to rub so as to remove the surface covering of scales, hair or other vestiture.

Deorsum: downward.

Dependent: hanging down.

Deplanate -us: see complanate.

Depressed: flattened down vertically; opposed to compressed.

Depressor: applied to a muscle that has for its function the depression of an organ or a part.

Deratoptera: = Orthoptera.

Dermal: relating to the skin or outer covering.

Dermal glands: hypodermal unicellular glands which secrete wax, setae, spines, etc.

Dermaptera: see Dermatoptera.

Dermatoptera: skin-winged: an ordinal term applied to insects with elytriform, abbreviated primaries beneath which the secondaries are folded transversely and fan-like: mouth mandibulate, prothorax free; abdomen forcipate; metamorphosis incomplete: the Forficulidae or earwigs.

Desectus: = truncatus.

Desideratum -ata: some thing or things needed or desired.

Destitutus: wanting; being without.

Determinate: with well-defined outlines or distinct limits: fixed: marked out.

Detonans: exploding: a sudden noise or a puff like an explosion.

Detritus: rubbed off; a surface partly denuded.

Deuterotoky: parthenogenetic reproduction when the progeny are male and female: see arrhenotoky and thelyotoky.

Deutocerebral segment: =antennal segment; q.v.

Deutocerebrum: the middle portion of the brain, formed by the ganglion of the 2d primary segment; also termed antennal or olfactory lobes from the parts it innervates.

Deutoplasm: the yolk or food plasm of an ovum.

Deutotergite: the secondary dorsal segment of the abdomen.

Dextrad: extending or directed toward the right.

Dextral: to the right of the median line.

Dextro-caudad: extends obliquely between dextrad and caudad.

Dextro-cephalad: extends obliquely between dextrad and cephalad.

Di: as a prefix, = two.

Diaphanous: semi-transparent; clear.

Diaphragm: any thin dividing membrane; that thin membrane separating the cavity containing the heart from the rest of the body.

Diarthrosis: any articulation that permits of motion.

Diastole: that regular expansion of the heart that draws the blood inward: see systole.

Dichaetae: a group of brachycerous Diptera with a proboscis consisting of two parts: Muscids, etc.

Dichoptic: Diptera; eyes separated by front: not contiguous: see holoptic.

Dichotomous: forked: dividing by pairs.

Dichromatism: the possession of two color varieties.

Dictyoptera: an ordinal term applied to the roaches: also more generally to the Orthoptera.

Didactyle -us: two-toed: with two tarsi of equal length.

Didymus: double: geminate.

Difformis: irregular in form or outline: not comparable; anomalous. Diffracted: bending in different directions.

Diffuse: spreading out; without distinct edge or margin.

Digestive tract: the alimentary canal as a whole: more specifically that portion behind the crop, in which assimilation takes place.

Digitate: finger-like, or divided into finger-like processes.

Digitiform: formed, shaped like or having the function of a finger.

Digitules: appendages on the feet of Coccidae; in Lecanium, four knobbed hairs.

Digitus: the terminal joint of the tarsus, bearing the claws: a small appendage attached to the lacinia of the maxilla; rarely present and probably tactile.

Digoneutism: the power to produce two broods in one season.

Dilatatus: Coleoptera a margin, when the sharp marginal edge extends beyond its usual limit: the base when the transverse diameter is much longer at one part.

Dilated: widened, expanded.

Dilation: an expansion or widening.

Dilute: thinned out: applied to color means weak or pale.

Dilution: much thinned out or diluted.

Dimera: forms with two-jointed tarsi: specifically applied to some groups of Homoptera.

Dimerous: having only two tarsal joints.

Dimidiate -us: halved; extending half way around; applied to elytra when they cover only half the abdomen.

Dimidius: of half length.

Dimorphic: occurring in two well-marked forms.

Dimorphism: a difference in form, color, etc, between individuals of the same species, characterizing two distinct types: may be seasonal, sexual or geographic.

Dioecious: with distinct sexes.

Dioptrate: an ocellate spot with the pupil divided by a transverse line.

Dioptric: with a transversely divided ocellus.

Diploglossata: an ordinal term proposed for Hemimeridae, because of the supposed presence of a second labial segment.

Diplogangliata: applied to the Arthropods.

Diploptera: = diplopteryga; q.v.

Diplopteryga: Hymenoptera; wasps in which the wings are longitudinally folded when at rest.

Dipneumones: having two lungs (certain spiders).

Diptera: an ordinal term applied to insects having only one pair of wings (anterior): thorax agglutinate; mouth haustellate; transformations complete.

Dipterocecidium: a gall formed by a dipterous insect.

Dipterous: belonging to or having the characters of Diptera.

Direct: applied to metamorphosis = incomplete.

Directive coloration: directive marks or colors which tend to divert the attention of an enemy from more vital parts.

Disc: see disk.

Discal: on or relating to the disc of any surface or structure.

Discal area: of a wing applies especially to the more central portion, or that area covered by the discal cell.

Discal bristles: Diptera; are inserted on the middle of the abdominal segments before the hind margin.

Discal cell: Lepidoptera; the large or median cell extending from the base of the wing toward the center: = radial cell (Comst.): in Diptera (Will.) = 1st medial 2 (Comst.): Odonata; = discoidal areolets, q.v.: Trichoptera, the cell between the forks of the radial sector, and separated from the 2d apical cell by a cross-vein.

Discal patch: in some male Hesperidae the oblique streak of specialized black scales on the disc of the primaries.

Discal vein: Lepidoptera; the cross-vein closing the discal or median cell extends from radius 5 to media 1.

Disciform: formed or shaped like a disc.

Discocellular nervure or vein: Lepidoptera; = discal vein, q.v.

Discoidal: relating to the disc, or middle = discal.

Discoidal area: the middle area or field: Trichoptera; that area of the tegmina between the posterior or anal and the anterior or costal areas = d. field.

Discoidal areolets: Odonata; a varying number of rows of cells on the outer side of the triangle between the short sector (M 4 of Comst.) and the upper sector of the triangle (Cu 1 of Comst.) = post-triangular cells := discal cells.

Discoidal cell: Hymenoptera (Norton) 1st medial 2, medial 3 and medial 4 (Comst.).

Discoidal field: see discoidal area.

Discoidal nervule: Lepidoptera; = media 1 (Comst.).

Discoidal triangle: Odonata - see triangle.

Discoidal vein: Diptera (Schiner), = media 2 (Comst.) anterior intercalary vein (Loew); Hymenopteran (Norton), = media 2 (Comst.), beyond the junction with the medial cross-vein: Trichoptera; the first and largest branch of the humeral vein.

Discoideous: =discoidal.

Discolored -orous: a different color from the surrounding, more or less contrasting; not concolorous.

Discota: insects in which development of the adults is from imaginal discs: see adiscota.

Discrete: distinctly separated.

Discs: the abdominal motor processes of coleopterous larve.

Discus: a disc; a somewhat flat circular part or area.

Disjoined or Disjointed: see disjunctus.

Disjunct: with head, thorax and abdomen separated by constrictions.

Disjunctus: separated; standing apart.

Disk: the central upper surface of any part; all the area within a margin; the central area of a wing: in Trichoptera, the obliquely ridged outer surface of hind femur in saltatoria.

Dislocated: a stria, band or line interrupted in continuity, when the tips of the interrupted parts are not in a right line with each other.

Disperses: with scattered markings, punctures or other small sculptures.

Disposed: arranged or laid out.

Dissepiment: a partition wall: applied to the forming septa separating the coelom-sacs in the embryo; also the thin envelope about the members in obtect pupae.

Dissilient: bursting open elastically.

Distad: toward the distal end.

Distal: that part of a joint farthest from the body.

Distant: remote from: standing considerably apart.

Distichous: applied to antennae when lateral processes originate at the apices of the joints and bend forward at acute angles to them.

Distiproboscis: the outer third of the proboscis in Muscid flies, bearing the labella.

Distychus: bipartite: separated into two parts.

Ditrocha: Hymenoptera; that series having the trochanter two-jointed.

Diurnae: day fliers: applied to butterflies.

Diurnal: such insects as are active or habitually fly by day only.

Divaricable: able to spread apart or divaricate.

Divaricate: straddling or spreading apart: when the wings are lapped at base and diverge behind: tarsal claws when arising at opposite sides of the joint and separating widely.

Divergent: spreading out from a common base; in Coleoptera, tarsal claws are divergent when they spread out only a little; divaricate when they separate widely.

Diverse: unequal: differing in size or shape: of various kinds.

Diverticulum -la: an oft-shoot from a vessel or from the alimentary canal usually blind or sac-like: applied to the caecal tubes or pouches: any extensions or evaginations of the hypodermic.

Dividens (vena): Trichoptera; 1st anal (Comst.).

Dog-ear marks: in bees: small, subtriangular marks of light color, just below the antennae (Cockerell).

Dolabriform: hatchet-shaped: compressed, with a prominent dilated keel and cylindrical base.

Dolioloides: applied to obtect or coarctate pupae.

Dominant: a character more constant and conspicuous than any other: a type or series occurring in large numbers both as to genera, species and individuals and in which differentiation is yet active.

Dorsad: extending or directed toward the upper side.

Dorsal: of or belonging to the upper surface: in Diptera, that face of the laterally extended legs visible from above.

Dorsal bristles: see dorso-central.

Dorsal diaphragm: the wings of the heart, or the very thin membrane upon which these muscles rest: = pericardial diaphragm, q.v.

Dorsal gland orifices: in Diaspinae, oval orifices arranged in more or less distinct rows on the surface of the pygidium, through which is discharged the material of which the dorsal scale is formed.

Dorsal glands: see last preceding title.

Dorsal line: in caterpillars, extends longitudinally on the middle of the back or dorsal.

Dorsal scale: that part of the covering scale of the Diaspinae that lies above the insect, as opposed to the ventral scale, which lies below.

Dorsal space: in slug-caterpillars is the area between the sub-dorsal ridges.

Dorsal vessel: the heart; q.v.

Dorsi-meson: the middle of the upper surface.

Dorso-alar region: Diptera; between the transverse suture and the scutellum on one side and the root of the wing and the dorso-central region on the other.

Dorso-central bristles: Diptera; two or four longitudinal rows on the inner part of the dorsal.

Dorso-central region: Diptera; bounded by two imaginary lines drawn from the scutellar bridges forward, and coinciding with a space free from bristles that exists on the outer side of the dorsal rows and is often occupied by a dorsal thoracic stripe.

Dorso-humeral region: Diptera; bounded by the anterior end of thorax and transverse suture on two sides and by the dorsopleural suture and dorsocentral region on the two others.

Dorsolum: the mesoscutum.

Dorsopleural suture: Diptera; the lateral suture between dorsal and pleurum from the humeri through the base of the wing: separates the mesonotum from the pleura.

Dorso-ventral: in a line from the upper to the lower surface.

Dorsulum: the mesonotum before the scutellum, with the wing sockets: also, specifically, the meso-scutellum.

Dorsum: the upper surface: in Coleoptera; often confined to meso- and metathorax: Odonata; includes mesepisterna and meso- and meta-thoracic terga: Diptera; upper surface of thorax, limited by the dorsopleural sutures laterally, the scutellum posteriorly and the neck anteriorly: Lepidoptera; the lower or inner margin of the wing.

Draw-thread: the silk-producing gland.

Drone: in Hymenoptera; the male bee.

Duct: a channel, tube or canal for carrying a secretion from a gland to the point of discharge.

Ductus ejaculatorius: the single duct or tube formed by the union of the vasa deferentia from each side, through which the seminal fluid is ejected into the vagina.

Dufour's gland: that gland, in Hymenoptera, that secretes the alkaline portion of the poison carried by the sting.

Duodenum: the chylific ventricle; also applied to the first section of the digestive tract just behind entrance of malpighian tubules.

Dupion: a cocoon spun by two silk-worms together; also the coarse silk from such a cocoon.

Duplicate -us: double.

Duplicate-pectinate: having the branches of a bipectinated antenna alternately long and short.

Duple: double, or twice.

Durus: hard.

Dusky: somewhat darkened; pale fuscous.

E

E: as prefix, is privative and means without.

Ears: organs of hearing, as on the first tibiae or on the first abdominal segment of some Trichoptera.

Ebenine: black like ebony.

Eburneous: ivory white.

Ecalcaratus: without a spur.

Ecaudate: without tails or tail-like processes: usually applied to wings : = excaudate.

Ecdysis: the process of casting the skin; moulting.

Echinate: set with prickles.

Ecology: the science of the relation of organisms to each other and to their surroundings: = ethology. {*Scanner's comment: Ethology nowadays refers to studies in animal behaviour, not directly to ecology*.}

Ectad: extending outwardly from within.

Ectal: belonging or relating to the outer surface.

Ectoblast: the outer wall of a cell; the ectoderm or epiblast.

Ectoderm: the outer layer of skin: the outer layer of the blastoderm, giving rise to the nervous system and to epithelial structures of the body surface.

Ectognathus: see ectotrophous.

Ectoskeletal: referring to the outside or exoskeleton.

Ectotrachea: the outer surface or layer of the trachea.

Ectotrophous: with mouth parts free; not buried in the head: see entrotrophus.

Edematus: dull translucent white.

Edentate -ulous: without teeth.

Edentula: those having no teeth.

Efferent: carrying outward or away from the centre.

Effluvium: a foul or unpleasant smell or emanation.

Effected: somewhat angularly bent outward.

Egg: a simple cell, capable of fertilization, containing the germ, the food-yolk necessary for its nutriment, and a covering membrane: a single ovum or cell from an ovary: the first stage of the insect.

Egg-burster: a projecting point on the head or other part of an embryo, used in breaking the shell when hatching.

Egg-calyx: the enlarged portion of the oviduct at the opening of the ovarian tubes, into which the egg is received before its entrance into the vagina.

Egg-case: the case or covering prepared or secreted by an insect to contain or hold together the egg-mass as a whole: see oötheca.

Egg-guide: Orthoptera; two small pointed prolongations of the ventral portion of the 8th abdominal segment, between upper and lower valves, used in oviposition.

Egg-pouch: see oötheca.

Egg-pod: applied to the egg-mass of grasshoppers.

Egg-tube: see ovarian tube.

Ejaculatory duct: see ductus ejaculatorius.

Elastic: a part which has a degree of flexibility throughout.

Elate -us: see elevatus.

Elater: the spring or forked tail of Podurids.

Eleutherata: all forms with free, separated maxillae; later, and more specifically, the Coleoptera.

Elevate -us: a part higher than its surroundings.

Elinguata: without a tongue: forms in which the maxillae are connate with the labium: see synista.

Ellipsoidal: see elliptical.

Elliptical: oblong-oval, the ends equally rounded, together forming an even ellipsoid.

Elongata -ate: drawn out; lengthened; much longer than wide.

Elutus: with scarcely distinct markings.

Elytra: the anterior leathery or chitinous wings of beetles, serving as coverings to the secondaries, commonly meeting in a straight line down the middle of dorsum in repose: also applied to the tegmina in Orthoptera.

Elytral ligula: a tongue-like process on the inner face of the side margin of elytra, to perfect the union with the ventral segments: e.g. in Dytiscidae.

Elytriform: shaped or appearing like an elytron.

Elytrin: = chitin, q.v.

Elytron: singular of elytra; q.v.

Elytroptera: see Coleoptera.

Emandibulata: that series of insects in which there are no functional mandibles in any stage.

Emandibulate: lacking functional mandibles; e.g. butterflies and moths, and applied in any stage.

Emarginate: notched: with an obtuse, rounded or quadrate section cut from a margin.

Embolium: Heteroptera; the narrow sclerite extending along the anterior margin of the hemelytra, from base to cuneus or membrane: the lobes on each side of the prothorax: the special enlargement at the base of the primaries which fits into a cavity in which the wing is moved.

Embossed: ornamented with raised figures.

Embryo: the young animal before leaving the body of the parent or before emerging from the egg.

Embryonic: found in, or relating to the embryo; in an undeveloped state or condition.

Emmet: an ant.

Empodium: Diptera; the small process between the pulvilli: in Coleoptera; the bifid pseudotarsi between the claws: used also as = pulvillus; and see arolium, onychium, palmula, paronychium, plantula, pseudonychium and pulvillus.

Enarthrosis: an articulation like a ball and socket joint.

Encephalon -um: the brain, or that part of the head containing it.

Encircled: ringed; margined round about.

Endemic: occurring normally where found: native, not introduced.

Endocardium: the inner lining membrane of the heart.

Endochorium: the layer of the allantois that lines the chorium; the inner layer of the chorium.

Endocranium: the inner surface of the cranium.

Endoderm: the inner layer of the blastoderm in the embryo, giving origin to the mid-intestine and other visceral organs: see entoderm.

Endolabium: the inner or mouth surface of the labium: the hypopharynx when that is well developed.

Endomesoderm: the inner layer formed by an invagination of the middle portion of the primitive band of the embryo, and from which the endoderm and mesoderm are subsequently differentiated.

Endophytic: living within plant or tree tissue, as borers or miners.

Endoskeletal: relating or referring to the endoskeleton.

Endoskeleton: applied to those chitinous processes extending inward into the body cavity from the body wall and serving as attachments for muscles.

Endosternite: that part of the apodeme arising from the intersternal membrane.

Endothorax: the internal framework or processes of the thorax.

Endotoky: is applied to that form of reproduction where the eggs are developed within the body of the mother; see exotoky.

Endotrachea: the inner surface or lining of the trachea: see intima.

Enervis: applied to wings without veins of any kind.

Engraved: see exsculptus.

Ensiform: sword-shaped: two-edged, large at base and tapering to the point: see anceps.

Entad: extending inwardly from without.

Ental: referring to the centre of the body cavity.

Enteric: relating to the digestive canal or enteron.

Enteron: the digestive canal as a whole; a general term.

Entire: with an even unbroken margin: said of wings when they are not divided or cut into.

Entoderm: the innermost germ layer of the embryo, from which are derived the epithelium of the alimentary canal and accessory structures: = endoderm and hypoblast.

Entognathous: see entotrophous.

Entoloma: the inner margin of the wings.

Entomogenous: growing in or on an insect: e.g. fungi.

Entomography: the description of an insect or of its life history.

Entomolin: = chitin, q.v.

Entomologist: one who collects and studies insects.

Entomology: that branch of Zoology that deals with insects and, specifically, the Hexapods.

Entomophagous: feeding upon insects: specifically applied to those wasps that feed their young with larvae, etc.

Entomophilous: insect-loving: applied to plants especially adapted for pollination by insects.

Entomophytous: referring to plants produced in or on an insect: see Entomogenous.

Entomosis: a disease caused by a parasitic insect.

Entomotaxy: the preservation and preparation of insects for study.

Entomotomy: that science which deals with internal structure of insects.

Entomotomist: a student of insect structure.

Entosternum: the internal processes from the sternum.

Entothorax: applied to the apodemes or processes extending inwardly from the sternal sclerites: see apophysis.

Entotrophous: with the mouth parts buried in the head:= entognathous: see ectotrophous.

Entozoa: those animals that live within the body of others.

Environment: the sum of the influences surrounding or acting upon an organism.

Enzyme: a ferment secreted by a cell or a gland.

Epalpate: having no palpi.

Ephebic: referring to the winged, adult stage.

Ephemerida: May-flies: an ordinal term used for insects with net- veined wings, held vertically when at rest, not folded; mouth mandibulate, not functionally developed: thorax loosely agglutinated; abdomen with anal filaments: metamorphosis incomplete.

Ephemeroptera: briefly winged: = ephemerida; q.v.

Epiblast: the outer germ layer of the embryo.

Epicranial: relating or pertaining to the epicranium.

Epicranial lobe: in caterpillars, the lateral, superior convex lobe of the head.

Epicranial plate: in some larvae a plate-like structure forming the epicranium.

Epicranial suture: the line of junction of the two procephalic lobes.

Epicranium: the upper part of the head from the front to the neck: often used to include front, vertex and genae:= calva.

Epideme: see articulatory epideme.

Epiderma -is: the cellular layer of the skin, underlying and secreting the cuticula: incorrectly applied to the outer skin or cuticle.

Epidermata: abnormal excrescences or outgrowths from the skin.

Epididymis: the convoluted efferent ducts, massed at the posterior part of the testes.

Epigastrium: the first entire ventral sclerite of the abdomen.

Epigenesis: the doctrine of growth from an undifferentiated germ, as opposed to preformation, which implies development from already existing rudiments.

Epigenetic: the period after the union of the male and female elements, during which organs are forming.

Epiglossa: = epipharynx; q.v.

Epiglottis: = epipharynx; q.v.

Epilabrum: a sclerite at each side of the labrum: specifically applied in myriapods.

Epilobe: of mentum in Carabidae, really corresponds to a partially divided ligula: a lateral appendage of a bilobed mentum.

Epimera -eron: the posterior lateral thoracic sclerites; usually small, narrow or triangular.

Epiopticon: the second ganglionic swelling of the optic tract: see opticon.

Epipharyngeal: belonging or relating to the epipharynx.

Epipharyngeal sclerites: in bees; a pair of strap-like pieces extending backward from the two sides of the base of epipharynx: see hypopharyngeal sclerites.

Epipharynx: an organ, probably of taste, attached to the inner surface of the labium and supposed to correspond to the palate of higher animals Epiglossa or epiglottis.

Epiphysis: a lappet-like process covering an excavation on the fore tibia of many Lepidoptera.

Epipleural: the deflexed or inflexed portions of the elytra, immediately beneath the edge: the inflexed portions of the pronotum are sometimes called prothoracic epipleura: as generally used, the term is incorrectly applied to the entire bent under margin of the elytra.

Epipleural fold: the raised lower edge of the epipleura: see hypomera.

Epiploön: see caul.

Epipygium: the dorsal arch of the last abdominal segment.

Episternites: the upper pair of corneous appendages forming the ovipositor in grasshoppers.

Episternum: the anterior and larger lateral thoracic sclerite between the sternum and notum.

Epistoma -is: the lower face between the mouth and eyes: that sclerite immediately behind or above the labrum, whether it be clypeus or an intermediate piece: in Diptera, that part of the face between the front and the labrum; the oral margin and an indefinite space immediately contiguous thereto and so = peristoma: in Odonata; = clypeus: = hypostoma.

Epithelium: the layer of cells which covers a surface or lines a cavity.

Epizoa: insects that infest the body surface of animals.

Epizoötic: living or parasitic on animals from the outside or on the surface.

Epomiae: the elevated margin of an oblique furrow in the propleurae for the reception of the front femora; Hymenoptera.

Epupillate: an ocellate spot included by a colored ring, but destitute of a pupil or central spot.

Equal: of the same length, size or shape: the superfices when they are without inequalities.

Equitant: laminated: folding one upon the other.

Erect: standing upright; not necessarily perpendicular.

Erectile: capable of being erected; applied to an appendage, a hair or other process, or to any tissue which may be distended and made rigid.

Erecto-patent: the wings of Hesperids when at rest; primaries erect, secondaries horizontal.

Eremochaetus: Diptera in which there is a general absence of bristles.

Ergatandrous: applied to ants with worker-like males.

Ergatogynous: applied to ants with worker-like females.

Ergatoid: sexually capable, wingless ants, resembling workers.

Ericeticolous: living in poor, sandy or gravelly places.

Eroded -sus: gnawed; a margin with irregular teeth and emarginations.

Eruca: broadly a larva: more specifically a caterpillar.

Eruciform: like a caterpillar in form or appearance.

Erucina: the caterpillar-like larvae of sawflies and the like.

Erucivorous: a feeder on caterpillars; said of parasites.

Erythrinus: red; nearly arterial blood-red: carmine, a little diluted.

Erythrinus: deep brick-red, tending to blood-red [vermilion with a little Indian red].

Escutcheon: the scutellum in Coleoptera.

Essential character: see specific character.

Ethology: see ecology.

Eucephalous: with a well-developed head, bearing the normal appendages: applied to certain dipterous larvae.

Eucone: a compound eye in which the individual ocelli have crystalline cones see acone.

Euorthoptera: the Orthoptera excluding the Dermaptera.

Euplexoptera: with beautifully folded wings: an ordinal term applied to the earwigs.

Eous or -eus: as a suffix, indicates the possession of the quality of the stem word: e.g. membraneous, like a membrane in texture.

Eutracheata: applied to articulates which, like the insects, have a well-developed tracheal system.

Evaginate: extruded by eversion; turned inside out when extruded.

Evagination: an extrusion formed by eversion or turning inside out.

Evanescent: disappearing; becoming gradually less.

Eversible: capable of being turned inside out.

Evident: easily seen or recognized.

Ex: prefix = A and E as privatives: also means from or out of.

Exarate -us: sulcated: sculptured.

Exarticulate: without distinct joints.

Exasperate -us: rough with irregular elevations.

Excalcarate: without spurs.

Excaudate: see ecaudate.

Excavate: with a depression that is not the segment of a circle.

Excentric: not in the centre; revolving or arranged about a point that is not central.

Excision: with a deep cut: a notch or other cut-out part.

Excrementaceous -titious: made up of or resembling excrement.

Excrescence: an outgrowth or elevation; usually abnormal.

Excretion: the act of getting rid of waste products: any material or substance produced by any secretory glands or structures and which is voided or otherwise sent out from them.

Excretory: those structures concerned in ridding the body of waste products.

Excurrent: attenuate, narrowly prolonged.

Excurved: curved outwards.

Ex larva: from or out of the larva: usually applied to specimens that have been bred from collected larvae.

Exochorion: that part of the chorion derived from the ectoderm: the outer layer of the chorion.

Exochorium: Heteroptera; a narrow marginal part of the hemelytra.

Exoderm: the outer skin or crust.

Exoloma: the apical margin of the wings.

Exophytic: relating to the outside of plant tissue.

Exoskeleton: the entire body wall, to the inner side of which muscles are attached.

Exotic: not a natives of the place where found: an introduced species: also any species occurring in any country outside of the limits of the country whose fauna is under consideration.

Exotoky: is applied to that form of reproduction where the eggs are developed outside of the body of the insect and without care by the mother see endotoky.

Ex ova: from or out of the egg: applied to specimens that have been bred from the egg stage.

Expanded: spread or flattened out: applied to Lepidoptera when set for the cabinet.

Expanse: the distance between the apices or other widest point of the wings when fully spread.

Expansio alarum: the wing stretch; see expanse.

Expiratory: relating to the act of expiration, when the abdomen is contracted and the air contained in the abdominal tracheae is presumably forced out of them.

Explanate: spread out and flattened; applied to a margin.

Explicate: unfolded; open; without folds or plica.

Exsculptate -tus: a surface with irregular, more or less longitudinal depressions, as if carved.

Exscutellate: having no scutel.

Exserted: protruded; projecting beyond the body or over a given point.

Exsertion: a protrusion: an extension of a line or other ornamentation beyond its ordinary course.

Extended: spread out: not lying one upon the other.

Extense: extended: expanded.

Extension plate: a structure at the base of the pulvillus whose function it is to extend it.

Extension sole: the pad-like pulvillus which may be extended by the extension plate through the pressure plate.

Extensor: that which extends or straightens out; applied to muscles.

Extenuate: to make or to become weak, thin or slender.

Exterior: the outside.

Exterior margin: the outer margin; sometimes used for costal margin.

External: belonging to or on the outside.

External area: Hymenoptera; the upper of the three cells or areas of the metanotum, between the median and lateral longitudinal carina, first lateral basal area.

External median area: Hymenoptera; the median of the three cells or areas between the median and lateral longitudinal carinae: = second lateral area.

Externomedial vein: in Hymenoptera (Norton) = radius (Comst.); in Orthoptera, = media (Comst.).

Externo-median nerve: the humeral and discoidal veins together.

Extra-ocular: remote from or beyond the eyes.

Extremity: the point most remote from base.

Extrorse -um: toward the outside.

Extrude: to turn or force out.

Exude: to ooze or flow slowly through minute openings.

Exuvia -iae -ium: the cast skin of a larval insect: in Diaspinae the larval skin when cast and incorporated in the scale.

Exuviate: to cast the skin: to moult.

Exuviation: the act of molting: the cast-off skin or exuvium.

Eyes: the organs of sight, composed of numerous facets, situated, one on each side of the head: the term is properly applied to compound eyes only but is sometimes used to designate also the simple eyes or ocelli.

F

Face or Facies: the upper or outer surface of any part or appendage: the front of the head between the compound eyes above the mouth to the vertex; usually applied to insects in which the head is -vertical: in bees extends between the eyes to the base of the antennae; in the Hymenoptera generally the area between antenne and clypeus: in flies the area between base of antennae, the oral margin, eyes and cheeks.

Facet: a small face or surface: one of the parts, areas or lens-like divisions of the compound eye.

Facial angle: the angle formed by the junction of the face and vertex.

Facial bristles: Diptera; a series on either side of the middle portion of the face, above the vibrissae, along the facialia.

Facial carinae: applied to both the carinae of the frontal costa and the accessory (lateral) carinae of the face; but usually restricted to the accessory carinae in Orthoptera.

Facial depression: = antennal fovea, q.v.

Facialium -ia: Diptera; that portion of the face between the lower part of the frontal fissure and the antennal fovea.

Facial quadrangle: in bees; the quadrangle bounded laterally by the eyes, above by a line between their summits and below by a similar line between their lowest points.

Facial ridges: Diptera; the elevated lateral borders of antennal grooves.

Facial tubercle: Diptera; a median convexity below middle of face.

Facies: the face: the general appearance or impression.

Falcate: sickle-shaped; convexly curved: a wing when deeply excavated below the apex so as to leave the latter acute and a little curved.

Falciform: curved like a sickle.

False legs: = spurious legs; = prolegs; q.v.

Family: a division of classification including a number of genera agreeing in one or a set of characters and so closely related that they are apparently descended from one stem: opinionative and indicated by the termination idae.

Farctus: fully filled.

Farinaceous: mealy: applied to powdery looking wings and surfaces.

Farinose: dotted with many single, flour-like spots: mealy.

Fascia: a transverse band or broad line; it is common when it crosses both wings or wing covers.

Fasciate: banded transversely.

Fascicle -ulus: a bundle of hair, threads or fibres.

Fasciculate: bundled; clustered as in a bundle; tufted: a surface when covered with bundles of long hair.

Fastigiate: flat-topped and of equal height: also applied to elytra that extend a little beyond the abdomen.

Fastigium: Orthoptera; the extreme point or front of vertex.

Fat-body: is the mass of oil or fat cells found, especially in larvae, surrounding the alimentary canal and some other internal organs.

Fatiscent: with cracks, crevices or openings.

Fauna: the assemblage of animals inhabiting a region or country.

Favose: with large deep holes, like the cells of a honeycomb.

Favus: a cell like that of a honeycomb.

Fecula: the excrement of insects.

Fecundation: the making fertile; as an egg by a spermatozoön.

Feeler: commonly applied to antennae; q.v.

Feelers: tactile organs: the term is usually applied to the antennae but sometimes to the palpi, as mouth-feelers.

Feet: the legs or organs of locomotion; one pair attached to each thoracic segment; composed of coxa, trochanter, femur, tibia and tarsus only; plural of foot; q.v.

Female: designated by "O+" the astronomical sign for Venus: that sex in which the ova are developed. {*Scanner's comment: The sign for Venus being an orthogonal cross or plus sign hanging vertically below a circle.*}

Femina: the female, or belonging to that sex.

Femorate -us: with abnormal or unusually developed femora or thighs.

Femoro-tibial: pertaining to both femur and tibia or to the articulation between them.

Femur -ora: the thigh: usually the stoutest segment of the leg, articulated to the body through trochanter and coxa and bearing the tibia at its distal end: in Coccidae and quite commonly, the femur and trochanter are considered as one, for measuring purposes.

Fenestra: a window; a transparent glassy spot or mark; a pellucid mark in a vein: a small, pale, membranous area at the base of the antenna in roaches.

Fenestrate: with transparent or window-like naked spots as in the wings of some Lepidoptera.

Fenestrate membrane: of the compound eye is at the base of the ommatidia, at their junction with the optic nerve; see retina.

Ferreous -eus: the metallic gray of polished iron.

Ferrugineous -ous, -eus, -osus: rusty red-brown [Dragon's blood, but brighter].

Ferrugino-testaeeous: a rusty yellow brown: a mixture of rusty red with dull yellow brown.

Fertilization: takes place when a spermatozoön enters through the micropyle of an ovum and unites with the cell nucleus: loosely applied like copulation or to its completion.

Festivus: variegated with bright colors.

Festooned: arranged in loops as if hung from nails.

Fibre: a thread-like structure of any tissue.

Fibrilla: rod or sliver-like nerve elements, often grouped like a bundle of short threads.

Fibrin: a proteid compound making up a large part of the muscular tissue: also found in blood and other body liquids.

Fibrinogen: a proteid substance of the blood and other body fluids, concerned in the production of fibrin.

Fibroin: a chemical compound found in silk, cobwebs and the like.

Fifth longitudinal vein: Diptera (Will.); = media 3 (Comst.).

Filament: a thread: a long slender process of equal diameter throughout: an elongated appendage.

Filariasis: a disease caused by the presence of minute worms or Filaria, transmitted by mosquitoes. {*Scanner's comment: Nowadays it is known that many kinds of filariasis are transmitted by other species of flies, in particular Simuliidae and Tabanidae*}

Filate: Diptera; antennae that are simple, without lateral hair or dilation: thread-like.

Filator: the silk spinning structure of caterpillars.

File: the diagonal ridged vein near the base of the tegmina in crickets, used in stridulating: in general any structure wherever situated that serves the same purpose.

Filicornia: insects with thread-like antennae; e.g. in Coleopteran, the Carabidae.

Filiform: thread-like: slender and of equal diameter.

Filippi's glands: a pair of secondary glands, opening into the silk glands of caterpillars near their anterior end.

Fillet: a transverse, raised structure between the antennae in Lepidoptera.

Filose: ending in a thread-like process.

Fimbria: thick, ciliated hair at the termination of any part: fringes.

Fimbriate: a margin or process when set with a fringe of hair closely placed.

Finger: of maxilla, is the digitus, q.v.

First clypeus: see post clypeus.

First inner apical nervure: in Hymenoptera (Nort.); is cubitus 1, from media 4, to first anal (Comst.).

First lateral suture: Odonata; starts from beneath base of front wing behind humeral suture and meets it behind second coxa.

First longitudinal vein: in Diptera; - radius 1 (Comst.).

First submarginal cross-nervure: Hymenoptera; part of the media and the radio-medial cross vein (Comst.).

Fissate: divided or split: with fissures or cracks.

Fissile -is: cleft or divided; as the wings in plume-moths: also used for lamellate.

Fissiparous: applied to that form of asexual generation in which the parent divides; each part becoming a new individual.

Fissure: a crevice: a narrow longitudinal opening: a slit.

Fissus: cleft: longitudinally divided nearly to base.

Fistula: a slender tube: specifically applied to the channel formed by the union of the two parts of proboscis in Lepidoptera.

Fistular: like a slender, cylindrical tube.

Flabellate: with long flat processes folding like a fan.

Flabelliform: fan-shaped.

Flabellum: a fan: a leafed structure: the transparent lobe at the end of the glossa in bees: also used as = flagellum; q.v.

Flabs: the lobes at the tip of the dipterous mouth:= labella; q.v.

Flaccid: feeble: limber: lax.

Flagelliform: whip-like; applied to a process.

Flagellum: that part of the antenna beyond the pedicel: a whip or whip-like process: the tail-like process of a spermatozoön.

Flammate -eus: flaming or fiery red [vermilion intensified].

Flange: a projecting rim or edge.

Flank: the sides of the thorax: the pleura.

Flaring: widening out like the mouth of a trumpet.

Flavescent: somewhat yellow.

Flavid: yellowed: sulphur yellow.

Flavo-testaceous: light yellow-brown: almost luteous.

Flavous -us: sulphur yellow [gamboge].

Flavo-vixens: green verging upon yellow [apple green + chrome yellow].

Flex: to bend: to curve back.

Flexible: pliable; with elastic properties.

Flexile -is: capable of being bent at an angle without breaking: flexible.

Flexuous -ose: almost zig-zag, without acute angles but more acute at angles than undulating: differs from sinuate in being alternately bent and nearly straight.

Flexor: that which bends; applied to muscles.

Flocculus -i: a hairy or bristly appendage on the posterior coxa of some Hymenoptera.

Floccus: a tuft of wool or wool-like hair

Flosculiferous: species that bear a flosculus.

Flosculus: a small, tubular lunulate anal organ with a central style, in certain Fulgorids.

Fluviatile: inhabiting the margins of running streams.

Fly-blows: eggs or young maggots of flesh flies: meat is fly-blown when such eggs or larvae have been deposited on it.

Flying-hairs: very long slender surface hairs set in punctures.

Foetid glands: glandular structures from which a foul smelling liquid may be ejected.

Foliaceous: leaf-like, or resembling a leaf.

Folioles: leaf-like processes from a margin or protuberance.

Follicle: = cocoon, q.v.: a cellular sac or tube, as of a gland or ovary.

Folliculate: enclosed in a case, cocoon or follicle.

Food reservoir: Lepidoptera, a blind sac or diverticulum from the bind part of oesophagus lying in abdomen dorsal to the stomach.

Foot: the tarsus, q.v.; improperly used to = leg; but in the plural form refers to legs rather than tarsi: see feet.

Foot-shield: in caterpillars, the chitinous plate on outer side of abdominal feet.

Foot-stalk: of the maxilla, is the stipes.

Foramen: an opening in the body wall for the passage of a vessel or nerve: any opening at an apex: the opening of a cocoon.

Foramen magnum; the opening on the posterior surface of the head to give passage to those structures that extend from head to thorax occipital foramen.

Foramina: small openings in the body wall: in Orthoptera,; the auditory organs on the anterior tibiae.

Forceps: hook or pincer-like processes terminating the abdomen, like specialized appendages of ear-wigs: similar processes in the male, used as clasping organs in copulation.

Forcipate: bearing forceps or similar structures.

Forcipiform: having the form of forceps or pincers.

Fore: anterior.

Foregut: extends from the mouth to the end of gizzard; its epithelium being formed from the ectodermal invagination known as the stomodaeum.

Forehead: in Mallophaga, the head in front of the mandibles and antennae.

Fore-intestine: =foregut, q.v.

Forficate: = forcipate, q.v.

Forks: Trichoptera; forks of veins in apical part of wing, numbered 1, 2, 3, etc.

Form: applied to representatives of a species which differ from the normal or type, in some uniform character; it is seasonal if it occurs at a period different from the type; dimorphic if there is an alternation of generations or two color patterns occur; or sexual if the members of one sex differ uniformly from those of the other.

Formic: of, pertaining to or derived from ants.

Formicary: an ant's nest or ant-hill.

Fornicate: arched or vaulted: concave within, convex without.

Fossa -ae: = fossula; q.v.

Fossoria: burrowers: in Orthoptera, the mole crickets and allies; in Hymenoptera, the digging wasps.

Fossorial: formed for or with the habit of digging or burrowing.

Fossula -ae: a deep groove or sinus with sharp edges: specifically applied to grooves on the head or sides of prothorax in which the antennae are concealed.

Fossulate: a surface with oblong impressions.

Fossulet: an elongated, shallow groove.

Fourth longitudinal vein: Diptera (Will.), = media 2 (Comst.).

Fovea, Foveola -ae: a shallow depression with well-marked sides: a pit.

Foveate: with foveae or pit-like depressions.

Foveolate: with shallow cavities like a honey-comb.

Fractus: broken: also applied to a geniculate antenna.

Fragile: easily breakable: thin and brittle.

Frass: the excrement; usually the excreted pellets of caterpillars.

Free: unrestricted in movement: not firmly joined with or united to any other part: said of pupae when all the parts and appendages are separately encased as in Coleopteran.

Frenatae: that series of Lepidoptera in which a more or less well- marked frenulum occurs.

Frenate: having a frenulum.

Frenulum: the spine, simple in males, compound in females, arising from the base of secondaries in many Lepidoptera, whose function it is to unite the wings in flight: in Cicada the triangular lateral piece on the mesonotum which connects with the trochlea: the anal area of secondaries and thus = tendo, q.v.

Frenulum hook: in the males of frenate Lepidoptera, a hook or fold into which the frenulum is fitted.

Frenum: that which holds things together: a lunate or triangular portion at the inner and hinder base of the wing in Odonata and Trichoptera; see tendo.

Fringe -es: an edging of hair, scales or other processes extending well beyond the margin and usually of even length: in Lepidoptera, fringes occur on the outer margins of all wings and consist of scales or hair projecting beyond the wing membrane.

Frog: the articular pan, - q.v.

Frons: = front; q.v.

Front: the anterior portion of head between base of antennae and below ocelli: in Homoptera, the vertical median area of face.

Frontal: referring to the front of head or anterior aspect of any part.

Frontal costa: Orthoptera, a prominent vertical ridge of bead which may be median or lateral: see median carina and lateral carina.

Frontal fastigium: in Orthoptera, that process of the face extending dorsad between the antennae and meeting or nearly meeting the fastigium of the vertex in Tettigidae.

Frontal fissure: Diptera; the impressed line extending from the frontal lunule to the border of the mouth.

Frontal lobes: in Psyllidae, two lobes or swellings more or less completely divided by a suture in which an ocellus is situated.

Frontal lunule: Diptera; an oval or crescentic space above the base of antennae in Cyclorrhapha, bounded by the frontal suture.

Frontal processes: Diptera; = antennal process, q.v.

Frontal ridge: in Coleopteran; a sharp ridge on the dorsal margin of the eye, extending forward.

Frontal stripe: Diptera; the middle of the front when membranous or discolored: = vitta frontalis.

Frontal suture: Diptera; separates the frontal lunule from that part of the head above it: in Coleopteran; = clypeal suture.

Frontal tubercles: in certain Aphids, are raised structures upon which the antennae are placed.

Frontal triangle: Diptera; the triangular space in males, between the eyes below, limited by a line drawn through base of antennae.

Frontal vesicle: in Odonata; that elevated area on the vertex upon which the ocelli are situated.

Fronto-orbital bristles: in Diptera; are placed on each side of the front, just below the vertical bristles.

Fugitive: soon disappearing; not permanent.

Fulcrant: the trochanter when continued along the femur, as in Carabids.

Fulcrum: the chitinous envelope at the base of mouth in Diptera and Hymenoptera, covering the beginning of the oesophagus: any structure that serves as a support to another..

Fulgidus: shining.

Fuliginous -osus: sooty or smoky brown [Van Dyke brown + a little black].

Fulvo-aeneous: brazen, with a touch of brownish yellow [brown pink].

Fulvous -us: tawny; light brown with much yellow; nearly orange [pale cadmium yellow + Indian red].

Fumate -us: smoky gray [gray].

Fumose: smoky.

Function: the work or duty which a given part or organ normally performs.

Fungicolous: living in or on fungi.

Funicle: the joints between the scape and club in Funiculate antennae: a small cord: a slender stalk.

Funiculate: whip-like: long, slender, composed of many flexible joints.

Funicule: a small, cord-like structure; especially when sheathed.

Funiculus: the main tendon of abdomen: in Hymenoptera a slender ligament connecting the propodeum to petiole on its dorsal aspect.

Furca: a fork: the anal appendage used for leaping in Thysanura; see furcula: the forked ental processes of the sternum.

Furcal orifice: see sternal into orifice.

Furcate: forked; divided nto two approximately equal divisions.

Furcula: a forked process: an osmaterium {*Scanner's comment: sic. See comment under "osmaterium".*}: in Collembola the spring or saltatory appendage borne by the fourth abdominal segment: in Orthoptera, a pair of backwardly directed appendages which overlie in a more or less forked position the base of the supra-anal plate.

Furred: covered with dense hair resembling fur.

Fuscescent: becoming brown; with a brown shading.

Fusco-ferruginous: brownish rust red.

Fuseo-piceous: pitch black with a brown tinge or admixture.

Fuseo-rufous: red-brown, approaching liver brown.

Fuseo-testaceous: dull reddish brown [brown ocher].

Fuscous -us: dark brown, approaching black; a plain mixture of black and red [crimson lake + black].

Fused: run together: applied when two normally separated markings become confluent and have a common outline.

Fusiform: spindle-shaped: tapering gradually to each end.

Fusulus: = spinneret, q.v.

G

Gales: the outer lobe of the maxilla, usually two-jointed, often hood-like, subject to great modifications in Hymenoptera and Diptera, and forms the coiled tongue in Lepidoptera.

Galeotheca: that part of the pupal case that covers the gales. Gall: an abnormal swelling or excrescence on a plant, produced by an insect: = cecidium.

Gallicolous: dwellers in galls, whether as producers or inquilines.

Gallivorous: feeding upon galls or gall tissue.

Gamogenesis: reproduction through fertilization: see agamogenesis.

Ganglion -ia: a nerve centre composed of a cell mass and fibres: the white disc-like bodies connected by a double cord, lying above the ventral surface within the body and forming the centre of the nervous system.

Gasterotheca: that part of the pupa case that covers the abdomen.

Gastric: of or belonging to the belly or to the stomach.

Gastric caeca: = caecum; q.v.

Gastro-coeli: a pair of usually transverse lateral pits near the base of the second abdominal tergite in some Hymenoptera.

Gastro-ileal folds: occur in some insects at the junction of the chylific ventricle with the ileum and serve as a valve.

Gastrula: that embryonic stage resembling a sac, with an outer layer of epiblastic cells and an inner layer of hypoblastic cells.

Gastrulation: the process of forming a gastrula.

Gathering hairs: the soft, flattened, often hooked hairs on the tongue of bees and other Hymenoptera; = hooked hairs.

Gelatinous: of a jelly-like texture or consistency: viscid.

Geminate: arranged in pairs composed of two similar parts: doubled.

Gemmate -us: marked with metallic or bright colored spots.

Gemmiparous: applied to that form of asexual reproduction where new individuals arise as buds from the germ body of the parent.

Gena -ae: the cheeks; includes that portion of the head on each side below the eyes, and extends to the gular suture: in Odonata the area between the eyes and clypeus and mouth parts: in Diptera the space between the lower border of the eye and oral margin, merging into face at front and limited by the occipital margin behind.

Genal bristles: Diptera; are on the cheeks near lower corner of eye.

Generalized: primitive: containing in combination characters that are separated and specialized in other forms.

Generation: used as the equivalent of brood; q.v.

Genicular arc: Orthoptera; a curved dark marking on the posterior knee-joint.

Geniculate: knee jointed: abruptly bent in an obtuse angle.

Geniculum: a little knee or bend.

Genital armature: all the processes concerned in copulation.

Genital hamule: a little hook or plate covering the anal cavity of the male: the supra-anal or genital hook: in Lepidoptera, the uncut: in Odonata, in the plural, one or two pairs of lateral processes of the male genitalia on the ventral surface of the second abdominal segment.

Genital hook: = genital hamule.

Genitalia: the external organs of generation with all appendages.

Genital lobes: in Odonata, a pair of-backward and downwardly directed processes from the 2d abdominal segment, between which the vesicle of the penis lies.

Genital papilla: in some Smynthurids, a tubercular elevation upon which the genital aperture opens.

Genital spike: the sheath of penis which, in male Diaspinae takes the form of a long mucronate spike.

Genital tuft: in Lepidoptera; an expansible tuft of fine hair believed to be scent-producing.

Genital valve: Odonata; a chitinous piece on each side of the ovipositor, derived from the sternum of abdominal segment 9: probably = outer pair of gonapophyses.

Genoholotype: the species on which a genus is founded, whether unique or one of a series, specifically named as generic type by the author.

Genolectotype: the one species of a series selected as the type of the genus in which the describer of the genus placed it, subsequent to the description.

Genosyntype: one of a series of species upon which a genus is founded, no one species being mentioned as type.

Genus: knee; the joint between femur and tibia.

Genus: an assemblage of species agreeing in some one character or series of characters; usually considered as arbitrary and opinionative, though some consider it a natural assemblage.

Geometrid: larva which, when walking, alternately elevate and straighten the middle of their body: opposed to rectigrade; q.v.

Geodephagous: = adephagous; q.v.

Geodromica: terrestrial Heteroptera in which the antennae are not concealed.

Geophilous: living on the ground: species that live on the surface or come freely into contact with it.

Germanium: an ovary: that portion of an ovarian tube containing the cell elements.

Germ-ball: reproductive cells in larvae from which, exceptionally, young may develop as buds.

Germ-band or Germinal band: that portion of a young embryo which is to become the future insect, when it is in the form of a band or strap and may or may not show the division into the future segments.

Gerontogeic: belonging to the old world: see neogeic.

Gibba: a rounded protuberance or prominence.

Gibbous: hump-backed; protuberant: said of a macula when it resembles a moon more than half full.

Gibbus: when the whole surface forms a hump or obtuse cone.

Gills: respiratory structures which function in water; distinguished as true or blood gills where contained blood conveys the absorbed oxygen from the gill to the tissues, and as tracheal gills when this conveyance is by contained tracheae.

Gilvus: = flavus; q.v.

Ginglymus: a hinge joint that permits flexion in one plane.

Gizzard: a pouch-like structure between the crop and chylific ventricle furnished with chitinous teeth or plates, in which the food is prepared for the digestive juices by grinding or merely sifting = cardia.

Glaber-rous: smooth; free from all vestiture.

Gland: a cellular sac which separates or secretes from the blood specific portions to produce characteristic products - e.g. wax, saliva, silk, etc.

Gland-bearing prominence: in Diaspinae a prominence on the margin, bearing a gland opening on the dorsal surface.

Gland orifice: in Coccidae, the external opening through which a gland pours its secretions.

Gland spines: in Coccidae; spiny appendages, each of which is supplied with a single gland whose opening is at the tip.

Glandular: having the character or function of a gland: used as descriptive of specialized hairs, spines or other processes.

Glassy: transparent; glass-like in appearance.

Glaucus: shining sea-green: whitish blue inclining to gray lavender.

Globose: formed like a globe or sphere.

Globulin: an albumenoid proteid compound formed in the blood of insects.

Glochis: a barbed point.

Glomerate: congregated or massed together.

Glossa: the inner lobe of second maxilla, corresponding to the lacing of first maxilla: loosely used as a synonym for tongue: especially applied to the coiled structure of the Lepidoptera; see also ligula.

Glossarium: Diptera; the labrum-epipharynx; q.v.

Glossata: a Fabrician term for Lepidoptera.

Glossate: furnished with a spiral tongue.

Glossotheca: that part of the pupa which covers the tongue.

Glutinose -ous: slimy; viscid.

Gnathal: relating or pertaining to the jaws.

Gnathite: a jaw or jaw-like appendage; in the plural, the mouth parts.

Gnathochilarium: a plate formed by the labial structures.

Gnathopoda: the arthropods: the first pair of legs; especially applied in crustaceans: mouth feet.

Goffered: a surface with regular impressions, closely set, and separated by narrow ridges: reticulated.

Gonapophyses: three pairs of processes in the Orthoptera, one arising from the eighth and two from the ninth abdominal segment op the ventral surface. They appear to = the rhabdites composing the ovipositor of other insects.

Gonyodon: a tooth-like articulated process at the apex of the femur in some Noctuidae.

Gonytheca: articulating surface of femur to which the tibia is joined.

Gorgeret: the barbed sting of the honey bee.

Gracile: slender; graceful.

Gradate -vim: one grade or step at a time: to arrange in a series: to blend so as to merge one into the other - e.g. colors.

Gradate veins: a transverse series of veins, each before or beyond the next.

Grammineus: grass-green [apple green].

Granose: like a string of beads; moniliform.

Granulated: covered with small grains.

Granule: a little grain or grain-like elevation.

Granulose: roughened with granules or made up of distinct grains.

Gregarious: living in societies or communities; but not social.

Grège: raw silk, including the gummy outer layer, as spun by a caterpillar.

Grès: the gummy layer surrounding the silk thread spun by a caterpillar.

Gressorious -vial: with legs fitted for walking: in Lepidoptera; the anterior legs aborted, the others fitted for walking.

Griscent: ashen gray.

Griseus: light gray; a mixture of white and black [gray].

Group: a division of classification used indefinitely for a series of allied species, genera or larger assemblages.

Grouped glands: see circumgenital glands.

Grub: an insect larva: a term loosely applied, but more specifically to larvae of Coleopteran and Hymenoptera.

Guanin: a white amorphous compound which occurs in the transparent areas of some wings, giving a milky tinge, and is also found in the photogenic organs of Lampyridae: an excretory substance, composition C5H5N5O (von Furth).

Guest: applied to those insects that live in nests or dwelling places of other species, not necessarily at the expense of the host.

Gula: the throat: that sclerite forming the central portion of the head beneath, extending from the submental to the posterior margin, and laterally bounded by the genae.

Gular peduncle: in Coleopteran = submental peduncle.

Gular suture: the line of division between the gulag or throat and the gene or cheeks.

Gulf strip: see semitropical or gulf strip.

Gullet: = oesophagus; q.v.

Gulo-mental: includes the region covered by the gulag and mentum.

Gustatory: elating to the sense of taste.

Gutta: a light spot on a dark ground.

Guttate: with light spots or drops on a dark ground.

Gymnocerata: insects with freely movable, conspicuous antennae: see cryptocerata.

Gymnogastra: Hymenoptera; species in which the venter is visible: see cryptogastra.

Gymnoptera: species with membranous wings not covered with scales.

Gynandromorphic: when an individual of one sex exhibits on one lateral half the organic characters of the other, more or less completely.

Gyri-cerebrales: lobes of the oesophageal ganglion of the embryo, connected with the primary lobe: = stalked bodies.

H

Habena: a fascia on the thorax.

Habit -us: the port or aspect: used to express a resemblance in general appearance.

Habitat; abbreviated Hab.: the region or place which an insect inhabits or where it was taken.

Haemoglobin: the coloring matter of blood which serves to carry oxygen.

Haemolymph: the watery blood or lymph-like nutritive fluid of the lower invertebrates.

Haemoxanthine: a dissolved albuminoid in the insect blood, which has both a respiratory and nutritive function.

Hair: a slender, flexible filament of equal diameter.

Hairy: covered or clothed with hair.

Halophilous: species living in salt marshes, or near the sea.

Halterata: the Diptera.

Halteres: the poisers or balancers: capitate movable filaments in Diptera, situated one on each side of the thorax and representing rudimentary hind wings.

Halteriptera: the Diptera.

Hamule -us: furnished with hooks, or bent like a hook.

Hammock: the hammock-like covering of the caterpillars of certain moths. Hamule: a little hook.

Hamuli: Odonata; one or two pairs of hooked processes projecting from the ventral surface of the 2d abdominal segment of the male; usually termed genital hamules: in Hymenoptera; minute hooks on the anterior margin of secondaries used to unite them in flight with the inner margin of primaries: in tree crickets, hook-like processes of the male genitalia.

Hamus: Lepidoptera; a hook or loop attached to the under side of costal margin of primaries near base, to receive the frenulum of male moths.

Harpago -ones: the inner basal lobes of the clasping organs of d culicids also, more generally = harpes.

Harpes: the lateral pieces of the male genitalia in Lepidoptera, used as clasping organs: also applied to the corneous hooks often borne by these lateral pieces, which are then termed valves; see clasper: in culicids an articulated process, sometimes jointed, at the base of inner side of side-piece, below and exterior to the harpagones.

Hastate: halbert-shaped: excavated at base and sides but with spreading lobes or angles.

Hastiform: = hastate.

Hatched: closely marked with numerous short, transverse lines.

Hatching spines: = egg burster; q.v.

Haustellate: formed for sucking: applied chiefly to mouth structures.

Haustellum: a sucker: applied to that portion of the mouth of a sucking insect through which liquid food is drawn into the gullet.

Head: the first or anterior region of the insect body, articulated at its base to the thorax, bearing the mouth structures and antennae. It is now believed to be made up of seven primitive segments, named in order: 1, the ocular or protocerebral; 2, the antenna or deutocerebral; 3, second antenna or tritocerebral; 4, mandibular; 5, superlingual; 6, maxillary; 7, labial or 2d maxillary.

Head vesicle: in Diptera, = ptilinum, q.v.

Heart: the dorsal vessel or tubular structure divided into chambers, lying just beneath the dorsal, which serves to propel the blood and controls the circulation.

Heautotype: = autotype; q.v.

Helcodermatus: a surface with ulcer-like depressions: applied also to the boring or tearing spines of pupae.

Heliciform: in the form of a spiral snail shell: applied to the cases of some Trichoptera.

Helocerous: with clavate antennae.

Helvolus: tawny or dully reddish yellow.

Helvus: honey yellow [brown pink + chrome lemon].

Hemelytra: a modification of the anterior wings of Heteroptera, coriaceous at base, membranous at tip, not meeting in a straight line at the middle: more specifically applied to the corium; q.v.: also used for the tegmina of Orthoptera.

Hemi: as a prefix, means half.

Hemimeroptera: an obsolete term for Hemiptera.

Hemimetabolous: manifesting an incomplete metamorphosis, but with a marked difference between the stages: specifically the Ephemerida, Odonata and Perlidae. {*Scanner's comment: nowadays applied to far more orders, generally to those that undergo a marked metamorphosis, but without a pupal stage.*}

Hemiptera: half-winged: an ordinal term applied to insects in which the mouth parts consist of four lancets inclosed in a jointed beak or rostrum; metamorphosis incomplete: the primaries may be of uniform texture throughout (Homoptera) or may be thickened at base, membranous at tip (Heteroptera).

Hemispheric: like the half of a globe or sphere.

Hepaticolor: liver-brown [dragon's blood].

Hepatic pouches: applied to caeca pouches; q.v.

Herbivorous: feeding upon plant tissue: a leaf feeder.

Heremetabola: with slight or incomplete metamorphosis, but with a resting stage at the end of the nymph life; specifically the Cicadidae.

Hermaphrodite: an individual in which the characters of both sexes are combined.

Hetero: as a prefix, unequal; different from.

Heterocera: Lepidoptera in which the antenna are of any form other than clubbed at tip: opposed to Rhopalocera.

Heterochrome: of different color: applied to species in which there are two color forms of one sex, one of which is like (homoeochrome), the opposite sex, as in certain Odonata and Lepidoptera.

Heterochrony: an irregular development in point of time, a later stage becoming evident before one that is earlier in ordinary course.

Heterogamy: applied to those cases in which two sexual or a sexual and parthenogenetic generation alternate.

Heterogeneous: a mixture of different forms; abnormal.

Heterogeny: the alternation of sexual and parthenogenetic generations.

Heterogyna: the ants: referring to the different kinds of females, - queens and workers, - as distinguished from males.

Heteromera: Coleopteran in which the anterior and middle tarsi are 5- jointed and the posterior are 4-jointed.

Heteromerous: having an unequal number of tarsal joints on the feet.

Heterometabola: differing among themselves in metamorphosis; but not manifesting abrupt stages.

Heteromorphous: the metamorphosis complete, in abrupt stages, the larva unlike the adult.

Heteronomous: if two parts, compared with each other, are of different quality: differing in development or function.

Heteropalpi: palpi with a different number of joints in male and female, as in some Trichoptera.

Heteroptera: an ordinal term applied to that series of Hemiptera in which the anterior wings differ in texture from the posterior, and the different regions of primaries differ in texture.

Heteropterous: with wings of different texture in different parts.

Heterotypical: a genus, described from more than one species, these differing in structure,

Hexachaetous: Diptera in which the mouth structures have six piercing setae.

Hexanephric: with six kidneys, or structures serving as such.

Hexapoda: tracheate arthropods with head, thorax and abdomen distinct, and only six legs in the adult stage: the true insects.

Hexapodal -ous: provided with six feet.

Hians: gaping.

Hibernaculum: a tent or sheath made out of a leaf or other material in which a larva hides or hibernates.

Hibernate: to pass the winter in a dormant condition.

Hicks' bottles: {*Scanner's comment: sic*} flask-shaped pits or depressions in the antennae of bees and ants: supposed to be the organs of hearing.

Hind angle: in primaries of Lepidoptera, is that point where inner and outer margin meet: = anal angle of secondaries.

Hind-body: the abdomen.

Hind-gut: the intestinal canal from the end of chylific ventricle to the Anus, including the malpighian tubules and anal glands.

Hind-head: Mallophaga; that part of head behind mandibles and antennae.

Hind-intestine: = hind-gut.

Hinge: of maxilla = cardo; q.v.: the point of articulation of a movable joint.

Hips: the coxa; q.v.

Hirsute: clothed with long, strong hair; shaggy.

Hispid: bristly: sparsely set with short, stiff hair.

Histoblast: the morphological unit or cell characteristic of a particular tissue.

Histogenesis: the formation and development of tissue.

Histolysis: the degeneration and dissolution of organic tissue.

Hoary: covered with a fine, white, silvery pubescence: pruinose q.v.

Holometabolous: having a complete transformation; with egg, larval, pupal and adult stages distinctly separated.

Holopneustic: having many pairs of open stigmata.

Holoptic: Diptera in which the eyes of male are contiguous between vertex and antennae: see dichoptic.

Holosericeus: with short, dense, silky hair, giving a satiny lustre.

Holotype: the unique type: = type; q.v.

Homelytra: elytra of similar or equal substance.

Homo: prefix = the same; similar.

Homochronic heredity: inheritance at corresponding periods of life.

Homochronous: changes in an organism which appear in the offspring at the same age at which they did in the parent.

Homodynamous: serially homologous: homology of the metameres.

Homoeochromatism: applied when over a given region many butterflies tend to vary similarly as regards color.

Homoeochrome: of the same color: see heterochrome.

Homoeomerous: all feet with an equal number of tarsal joints: = isomerous.

Homoeonomous: of the same substance or texture.

Homoetype: = homotype; q.v.

Homogeneous: of the same kind or nature: similar in texture or parts.

Homogenous: similar in structure due to a community of descent.

Homologous: implies that organs are identical in general structure and origin, though they may have developed in different ways for special purposes: see analogous.

Homomorpha: insects in which the larvae resemble the adults.

Homonymous: pertaining to homology of parts arranged on a transverse axis similarly developed and of equal function.

Homonym: a name similar to or like another already used for a species in the same genus, or for a genus in the same kingdom: such names are paid to be preoccupied.

Homonymous: where the same name is applied to different conceptions.

Homophonous: words differently written but indistinguishable in sound, applied to different conceptions.

Homoplastic: implies that organs, similar in situation and purpose, are not structurally the same, or have not the same origin.

Homoptera: an ordinal term applied to those Hemiptera in which the primaries are of the same consistence throughout.

Homotenous: retaining the primitive form: applied to insects without or with an incomplete metamorphosis.

Homotype: is a specimen named by another than the author after comparison with the type.

Honey dew: a sweetish excretion produced by certain insects, notably Aphids and Coccids, and exuding from the surface of some galls.

Honey tubes: small tubes or tubercles on the abdomen of plant lice and other insects through which a sweetish liquid or honey dew is excreted siphonets; siphuncles; cornicles.

Hood: of the maxilla is the galena; q.v.: in Tingitidae the elevated portion of the prothorax, often covering the head.

Hooked hairs: = gathering hairs; q.v.

Horismology: see orismology.

Horizontal: said of wings when held parallel to the horizon.

Horn: a pointed chitinous process of the head: in the plural form applied to the antennae; q.v.

Host: the individual infested by or upon which a parasite grows: also applied to the maker of a cell or other structure in which guest flies or other insects take up their abode.

Hudsonian zone: is that part of the boreal region comprising the northern part of the great transcontinental coniferous forests. In the eastern United States restricted to the cold summits of the highest mountains, from northern New England to western North Carolina: in the west it covers the higher slopes of the Rocky and Sierra-Cascade systems.

Humeral: relating to the shoulder or humerus.

Humeral angle: in Lepidoptera, that angle of the wings at the base of costa, near the point of attachment to the body: in Coleopteran, the outer anterior angle of elytra: in Orthoptera, the obtusely rounded angle formed by the deflection of the sides of the pronotum from the dorsal.

Humeral bristles: in Diptera, are situated on the humeral callus.

Humeral callus: in Diptera, is a rounded callus forming the anterior superior angle of the mesothorax.

Humeral carina: in Coleoptera, an elevated ridge or keel on the outer anterior angle of elytra.

Humeral cross-vein: (Comst.); extends between the costa and subcosta close to base.

Humeralis: Coleopteran; when the elytral has an angulated projecting margin at base.

Humeral stripe: in Odonata, covers the humeral suture.

Humeral suture: in Odonata, runs from just in front the base of the fore-wing to the edge of the median coxa, separating the mesepisternum from the mesepimeron.

Humeral veins: in Lepidoptera, secondary veins on posterior wings of Lasiocampids, developed to strengthen the humeral angle.

Humerus: the shoulder: in Coleopteran; the basal exterior angle of elytra: in Diptera, the anterior superior angles of the mesothorax: in Orthoptera, the femur of the fore-leg: in Hymenoptera, applied to the subcostal vein in some groups.

Humid: applied to regions in which the normal rainfall is sufficient to produce ordinary farm crops without irrigation: see arid.

Hyacinthine: the purple blue of the hyacinth [between mauve and lilac].

Hyaline: vitreous: transparent or partially so.

Hyaloplasm: the clear, semi-fluid material between the meshes of the cell reticulum.

Hybrid: the progeny from the mating of two species.

Hydradephaga; -ous: applied to aquatic, predatory pentamerous beetles with filiform antennae: see adephagous.

Hydro: relating to water: a combining form used as a prefix.

Hydrolysis: the chemical decomposition of a compound by water, causing formation of a new compound.

Hydrophilous: applied to species living in low, damp places.

Hymen: a thin plane membrane serving as a partition.

Hymenoptera: membrane-winged: an ordinal term applied to insects with four membranous wings with few veins, the anterior usually larger than the posterior; mouth mandibulate; head free; thorax agglutinate, transformations complete.

Hyoid: having the form of the Greek upsilon, Y

Hypermetamorphosis: when an insect passes through more than the normal number of stages; the interpolated stages coming usually between the full-grown larva and adult.

Hyperparasite: is a form that is parasitic upon another parasite.

Hypertely: beyond the bounds of the useful: those forms whose resemblance to other objects is closer than needful, or without apparent object.

Hypertrigonal space: = supra-triangular space; q.v.

Hypertrophied: abnormally large or excessively developed.

Hypnody: lethargy; a condition similar to or identical with hibernation.

Hypertrophy: any abnormal enlargement or excessive development.

Hypoblast: = entoderm.

Hypocrateriform: salver-shaped.

Hypodactyle: the so-called labium of Hemiptera.

Hypoderm -is: the cellular layer which secretes the chitinous cuticula and in this sense = epidermis: specifically applied to the lining membrane of elytral and hemelytra.

Hypodermatic: of or concerning the hypodermic.

Hypodermic: under the skin.

Hypoglottis: the under surface of the tongue = hypoglottis.

Hypoglottis: a sclerite inserted between rectum and labium in many Coleopteran.

Hypognathous: having the mouth parts directed more or less vertically ventrad.

Hypographous: shaded; applied to a fascia that becomes gradually darker.

Hypomeron -a: in Coleopteran; the inflexed edge of the pronotum (pronotal hypomera); and the raised lower margin of the epipleural (elytral hypomera) (see epipleural) fold.

Hypopharyngeal: relating to the hypopharynx.

Hypopharyngeal sclerites: in bees, a pair of strap-like pieces along the hypopharynx to the mentum: see also epipharyngeal sclerites.

Hypopharynx: a sensitive and sensory structure on the upper surface of labium that serves as an organ of taste, or true tongue.

Hypopleura: in Diptera, the space over the middle and hind coxa, between the metapleura and pteropleura: the side of the metasternum: the mesepimeron of the mesothorax.

Hypoptere: = tegula; q.v..

Hypopygium: the anus: more specifically the lower plate of the anal opening: in Diptera, the male sexual organs and terminal segments of abdomen = propygium.

Hypostoma: in Diptera; that portion of the head included between antennae, eyes and mouth: in Hemiptera: the lower part of face.

Hypotenuses: in Odonata; the simple or broken cross-vein between media 4, and cubitus 1, forming outer boundary of triangle.

Hypotypes: includes specimens upon which supplementary descriptions are based: = apotypes.

I

Iceous or Icius: suffix; expresses a likeness or the possession of a character see aceus.

Icotypes: typical specimens which serve for purposes of identification, but have not been used in literature.

Idiotype: a specimen named by the author after comparison with the type, but not also a topotype.

Ignitus: fire-red [vermilion].

Ileo-colon: the anterior portion of the hind-gut, extending from the mid-gut to the rectum, when not distinctly differentiated into ileum and colon.

Ileum: the small intestine; begins at end of chylific ventricle at the point where malpighian tubules join, and extends to colon.

Imaginal: pertaining to the adult or imago.

Imaginal buds, cells, or discs: in forms with a complete metamorphosis are those embryonic cells around and from which the organs and appendages of the future imago develop.

Imago: the adult or sexually developed insect.

Imbricate: arranged or appearing like the scales on a fish or the shingles on a roof.

Immaculate: destitute of spots or marks.

Immarginate: without an elevated rim or margin.

Immersed -us: inserted, imbedded or hidden in.

Imponderable: that which cannot be weighed.

Impregnate-ed: to make or made fertile or pregnant: fertilized.

Impressed -us: a surface with shallow depressed areas or markings.

Impubis: without hair.

Inaequalis: unequal.

Inarticulate: not jointed or segmented.

Inaurate -us: golden yellow [pale cadmium yellow].

Ineanus: hoary.

Inch: the English and American standard of length in insect measurement: it is = 12 lines and = 25.4 mm.: usually expressed in units and hundredths, as 1.01.

Incised: notched or deeply cut into.

Incision: any cut into a margin or through a surface: the marginal slits or notches in Coccidae.

Incisure: an impressed line marking the junction of two segments: an incision.

Inclinate -us: leaning or inclining.

Inclusus: when one part is wholly or partially hidden in another.

Inconspicuous: not attracting attention or quickly noticeable.

Incrassated: thickened: rather suddenly swollen at some one point, especially near tip.

Incubate: to brood: to cause to develop; as an egg.

Incumbent: lying one over another: wings when they cover the dorsal horizontally.

Incunabulum: = folliculus and cocoon; q.v.

Incurved -ate: bowed or curved inwards.

Independent: in Lepidoptera; that vein of the wings that arises from the cross-vein closing the cell, and does not branch directly from any vein reaching the base: it is v. 5 of the numerical series in both wings and the media of Comstock.

Indeterminate: not defined nor well marked; obscure: of no constant form or shape.

Indigote: a very deep indigo blue.

Indirect: applied to metamorphosis = complete.

Indumentum: a covering of hairs, scales or tufts.

Indurated: hardened.

Indusium: the case made by an insect larva: a membranous layer of the embryo of Locustidae below the serosa.

Inequal: a surface with irregular elevations and depressions.

Inermis: unarmed: without spines or spurs.

Infericornia: Hemiptera; in which the antenna appear to be inserted well down on the sides of head; e.g. Lygaeidae.

Inferior: beneath, below or behind: a term of position.

Inferior appendage -es: in male Odonata the lower one or two of the terminal abdominal parts used to clasp the female in copulation.

Inferior wings: = hind wings or secondaries: q.v.

Infero-posterior: below and behind: refers to location.

Inflated: blown up; distended bladder-like.

Inflected: bent inward at an angle.

Inflexus: = inflected.

Infra: below or beneath: opposed to supra.

Infra-anal lobe: a thick, conical fleshy lobe, often ending in a chitinous point, situated beneath the vent in caterpillars.

Infra-cereal plates: in Orthoptera - generally inconspicuous paired plates which underlie in part the cerci and in part the lateral portion of the supra-anal plate.

Infra-clypeus: = ante-clypeus and rhinarium: q.v..

Infracted: abruptly bent inward, as if broken.

Infra-genital: below the genital opening or process.

Infra-marginal: situated below or behind the marginal cell.

Infra-median vein; in Orthoptera: = ulnar vein: q.v.

Infra-ocular: applied to the region below and between the eyes.

Infra-oesophageal: situated below oesophagus; see sub-oesophageal.

Infra-stigmatal: situated below the stigmata or spiracles.

Infringing: encroaching upon.

Infumated: clouded.

Infundibuliform: funnel-shaped.

Infuseated: smoky gray-brown, with a blackish tinge [Roman sepia].

Ingens: unusually large or disproportionate in size.

Ingluvies: the crop; q.v.

Inner lobe: of maxilla = lacinia: q.v.

Inner margin: the line extending along the lower or interior edge of the wing from the base to the hind or anal angle.

Innervate: to supply with nerves.

Innotatus: without markings.

Inocular: inserted in the inner margin of and partially or wholly surrounded by the eye.

Inquiline: a species living in a gall or other structure prepared by a different species, not as a parasite but as, a guest.

Inquiline: living as guests in the homes of others; as in galls.

Insect: a member of the class Insecta strictly limited.

Insecta: broadly defined, contains all articulates that are also tracheates and have the head free from the thorax; more strictly limited to those forms that have only three pairs of thoracic legs in the adult stage and a limited number of segments.

Insectary: a place or building where insects are bred and studied.

Insectivorous: feeding upon or devouring insects.

Insectologist: a student of insects: = entomologist.

Insectology: the science of insect study: = entomology.

Insertion: the point or place where a part is inserted: a part that is inserted: the act of inserting.

Insertus: a part that has its base set into another.

In situ: in its natural place or normal position.

Instar: the period or stage between molts in the larva, numbered to designate the various periods; e.g. the first instar is the stage between the egg and first molt, etc.: see stadium.

Institia: stria or furrows of equal width throughout.

Instrumenta cibaria: mouth parts of a mandibulate insect as a whole.

Instrumenta suctoria: mouth parts of a haustellate insect as a whole.

Integer: entire: applied to a margin without incisions.

Integument: the outer covering to the insect body.

Inter: between; among.

Inter-alar space: in Odonata; the terga of meson- and meta-thorax.

Interantennal: between the basal segments of antennae.

Inter-articular: the membranous tissue between joints or segments.

Intercalary -ies: additional or inserted between others; as a vein: plural; added or supplementary longitudinal wing reins: see under specific headings; i.e. anterior, etc.: in Ephemerides, certain longitudinal veins between the 8th (anal) and 9th (1st maxillary) and not branches of either: in Diptera, the anterior intercalary (Loew) = the discoidal, and the posterior intercalary = the cubitus 1 of Comstock: applied to an evanescent sclerite in the embryo between antenna and mandible; also termed premandibular.

Intercostal: between veins or costae; usually in the narrow grooves between veins in the costal region of a wing.

Intercostula: those small, vein-like structures between the normal veins, visible on a wing margin but lost toward the disc.

Intercoxal process: in Coleopteran; a median protrusion of the basal segment of abdomen between the hind coxae.

Intermediate: lying between others in position or possessing characters between two other forms.

Intermediate field: of termini is = discoidal field q.v.

Internal area: in Hymenoptera; the posterior of the three areas between median and lateral longitudinal carina on the metanotum third lateral area.

Internal cell: in Hymenoptera (Pack.) 2d anal (Comst.).

Internal triangle: in Odonata see triangle.

Internal veins: in Lepidoptera, from one to three in number, run free from base to outer margin near hind angle; never branched;1a to is in the numerical series: = anal veins (Comst.).

Interneural: between the nerves (or veins) of wings.

Interno-mandibular: applied to one of the pairs of salivary glands in bees, situated at the inner side of base of mandible.

Internomedian: in Orthoptera; = cubitus (Comst.); q.v.

Interocular: between the eyes.

Interplical: lying between folds; specifically applied to the alternate ridges and grooves in anal area of secondaries of Orthoptera.

Interposed sectors: in Odonata; the shorter longitudinal veins occurring in the wings of some species between the chief veins; = supplementary sectors. Interrupted: broken in continuity, but with the tips of the broken parts in a right line with each other.

Intersegmental: = interarticular; q.v.

Interspace: Coleopteran; the plane surface between elytral striae: Lepidoptera spaces between wing veins not included in closed cells.

Interspaceal: occurring in the interspaces between two wing veins or two elytral striae.

Interstice -tium: space between two lines, whether striate or punctate.

Interstitial line: the elevated ridge between two striae or series of punctures.

Interval: the space or time between two structures, sculptures or periods of development.

Interventricular: the inner valve between the chambers of the heart.

Interventricular valvule: of heart, lies in front of seluilunar valve.

Intervenular: in the space between two veins.

Intestinal caecum: that point of the large intestine in front of the junction with the small intestine.

Intestine: that part of the alimentary canal through which the food passes from the stomach, in which absorption is completed and the excretions are formed for expulsion.

Intima: the lining membrane of the trachea: see endotrachea.

Intorted: turned or twisted inwardly.

Intra-: within: between.

Intra-alar bristles: in Diptera; a row of two or three between the supra-alar and dorso-central groups.

Intracellular: occurring within the cell or in a cell.

Infra-humeral bristles: in Diptera calyptrata; occur immediately in front of the thoracic suture, between the humeral callus and the presutural depression.

Infra-ocular: situated within the eye, actually or apparently.

Intra-pulmonary: that method of respiration which does not involve movements of the outer body wall and is confined to the respiratory organs.

Intrauterine: applied to development, when the young hatch within the vagina of the mother.

Intricate: irregular: confused; applied to markings and sculpture.

Intromittent: used for throwing within.

Intromittent organ: the penis; q.v.

Introse -um: directed inward, toward the body.

Intrusus: seemingly impressed with a sharp point.

Intumescent: enlarged; swollen: expanded.

Invaginate: when a tubular or vesicular part is turned inward or retracted within the body wall.

Invagination: a pouch or sac formed by an infolding or indrawing of the outer surface.

Investitus: unclothed: a surface without scales or hair.

Involucrate: = involute.

Involucrum alarum in Dermaptera a flap of the metanotum.

Involute: spirally rolled inwardly.

Involuti: butterflies whose larvae live in a folded leaf; Hesperidae.

Iridescent: a surface which reflects the prismatic hues.

Iridicolor: any color so broken up as to reflect the prismatic hues.

Iris: the circle which, in an ocellate spot surrounds the pupil.

Irised: with rainbow colors.

Iris-pigment: = iris tapetum.

Iris tapetum: the pigment layer of the compound eye just below the crystalline cone.

Irregular: unequal, curved, bent or otherwise twisted or modified without order or symmetry, e.g. certain antennae.

Irrorate: marked with minute points; freckled.

Isabelline -us: pale yellow with some red and brown [chronic lemon with a little carmine and roman sepia].

Ischia: = pleura; q.v.

Iso-: equal.

Isolate: to separate out from others; occurring alone.

Isomers: that series of Coleoptera in which the tarsi have an equal number of joints on all feet.

Isomerous: with equal number of tarsal joints on all feet := homoeomerous.

Isomorphous: having the same form, appearance or construction.

Isopalpi: that series of Trichoptera in which the palpi of both sexes have the same number of joints.

Isoptera: equal winged: an ordinal term for insects with four, similar, net-veined wings; mouth mandibulate; thoracic rings similar, loosely jointed metamorphosis incomplete: the Termitidae. {*Scanner's note: In modern nomenclature the Isoptera constitute the order of all termites; the Termitidae are just one family within the Isoptera.*}

Isotypical: a genus described from more than one species, all of which are congeneric.

-itus: = -atus; q.v.

-ius: suffix; having the power or ability to.

J

Jabot: the crop; q.v.

Janthine: violet colored.

Jaw-capsule: contains the mouth structures in those dipterous larvae in which the head is differentiated.

Johnston's organ: a complex nervous structure in the basal joint of dipterous antennae.

Joint: a segment or part between two incisures: an articulation.

Jubate -us: fringed with long pendent hairs.

Juga: the lateral anterior lobes of the head of a Heteropteron; each side of the tylo.

Jugatae: that series of Lepidoptera in which there is a jugum instead of a frenulum to unite the wings in flight.

Jugular: of or pertaining to the throat.

Jugular sclerite: small sclerite in the membrane connecting the head with the thorax: see cervical sclerite.

Jugulum: that sclerite just behind the sub-mentum; =gula: that cavity of the posterior part of the head to which the neck is annexed: the lateral and under parts of the prothorax.

Jugum: in certain Lepidoptera and Trichoptera, a lobe or process at the base of primaries, overlapping secondaries and holding the two together in flight.

K

Katabolic: the destructive change from animal tissue to waste product: see anabolic.

Keel: an elevated ridge or carina.

Kermesinus: dark red, with much blue [purple lake].

Key: a tabular or other arrangement of species, genera or other classification according to characters that serve to identify them.

Kidney-shaped: like a kidney in outline; convex on one long side, concave on the other, the ends evenly and somewhat obtusely rounded: bean-shaped.

Knee: the point of junction of femur and tibia.

L

Labellum -a: the sensitive ridged tip of the mouth structures of certain Diptera: a prolongation of the labium covering the base of rostrum in Coleoptera and Hemiptera.

Labial: referring, pertaining or belonging to the labium.

Labial segment: the 7th segment of head = second maxillary segment.

Labial suture: is between labium and mentum.

Labiate: lip-like or having lip-like sutures.

Labipalp: a labial palpus.

Labis: the slender abdominal forceps in some Lepidoptera.

Labium: the lower lip: a compound structure which forms the floor of the mouth in mandibulate insects, behind the first maxilla and opposed to the labium; formed by a fusion in embryonic life of separate right and left maxilla-like halves: in some of its developments referred to as the tongue.

Labral suture: is between labrum and clypeus.

Labro-frontal lobes: of brain, = trito-cerebrum; q.v.

Labrum: the upper lip; covers the base of the mandible and forms the roof of the mouth.

Labrum-epipharynx: in the mouth of piercing Diptera is the central unpaired lancet.

Lac: a mixture of resin, wax and other substances produced by certain scale insects as a protective covering.

Lacer: a lappet; applied to a margin with irregular, broad and deep emarginations, leaving lappet-like intervals.

Lacerated: ragged; torn in appearance; see lacer.

Lacinia: the inner lobe of first maxilla, articulated to the stipes, bearing brushes of hair or spines: a blade: in Diptera, forms a flat lancet-like piercing structure and is never jointed.

Lacinia exteriores and interiores: in Apidae, the palpiger and paraglossa often used for the gales and lacinia of maxilla.

Laciniated: jagged; cut into irregular fragments.

Lacte: milk-white.

Lacteal: relating to milk; milky in appearance.

Lactescent: secreting or yielding a milky fluid.

Lacteous -eus: white, with a slight bluish tinge, like skim-milk.

Lacunae: irregular impressions or cavities: specifically the non-walled cavities of the body.

Lacunose: pitted; the surface covered with small cavities.

Laemodipodiform: like a laemodipod; similar to the larva of a walking stick.

Laete: bright.

Laevis -igatus: smooth, shining and without elevations: said of a surface.

Lamella: a thin plate or leaf-like process.

Lamellate: antennae with the club formed of closely opposed leaf-like surfaces, the concealed surfaces set with sensory pits.

Lamellicornia: those beetles in which the antennae terminate in a lamellate or leaf-like club.

Lamelliform: made up of or resembling leaves, blades or lamellae.

Lamina -ae: a chitinous plate or plates.

Lamina externa: the paraglossa.

Lamina interna: the ligula.

Lamina subgenitalis: the sub-genital plate; q.v.: in roaches the 7th ventral plate of females and 9th ventral plate of males.

Lamina supra-analis: = supra-anal plate.

Laminate: formed of thin, flat layers or leaves.

Laminato-carinate: with an elevated ridge or keel, formed of thin plates.

Laminiform: layer-like: having the appearance or made up of lamina.

Lana: wool: the long hair on the abdomen of some Lepidoptera.

Lanate -atus: woolly: covered with dense, fine, long hairs, so distinct that they may be separated.

Lanceolate: lance- or spear-shaped: oblong and tapering to the end.

Lanceolate cell: in Hymenoptera (ort.); - 2d anal (Comst.).

Lancet: indiscriminately applied to any piercing mouth structure.

Lanuginose -us: with long, curled hair dispersed over the surface: see crinitus.

Lanugo: slender single hairs.

Laparostict: that series of lamellicorn beetles in which the abdominal spiracles are situated on the connecting membrane between the dorsal and ventral rings.

Lapidicolous: living under deeply imbedded stones.

Larva: the second stage of insect development; comes from the egg or ovum, grows, and according to its kind, changes to a pupa or chrysalis or to an imago; bears various names in the different orders: see nymph; caterpillar slug; maggot; grub.

Larvarium: a tube or case made by a larva as a shelter or retreat.

Larvatae: asked; applied to coarctate and obtect pupae.

Larvina: a maggot: a dipterous larva without distinct head or legs.

Larvule: applied to early stages of Ephemerid larvae when they appear to have no developed respiratory, circulatory or nervous systems.

Lashed: eyes that have a more or less complete fringe of stiff hairs or bristles at the orbits.

Lasureus: a very dark blue [French blue with some black].

Laterad: toward the side and away from the median line.

Lateral: relating, pertaining or attached to the side.

Lateral areas: in Hymenoptera; on the metanotum, the three spaces between the median and lateral long carinae; the upper is the external or first lateral basal area; the second is the external or central lateral area; the third is the middle, internal, apical or third lateral area.

Lateral bristles: in Diptera; situated at or near the lateral margins of the abdominal segments.

Lateral carinae: in Orthoptera; on the head, extend downward from the front margin of the eyes: on prothorax extend along each lateral margin of the dorsum.

Lateral foveolae: in Orthoptera: foveate depressions on the margins of the vertex near the front border of the eye.

Lateral line: in caterpillars is at the margin of the dorsum between sub-dorsal and supra-stigmatal line.

Lateral lobe: of the labium in Odonata, corresponds to the paraglossa with palpiger and palpus (Gerstaecker) or, more probably, to the palpus alone (Butler).

Lateral lobes: the deflexed portions of pronotum that cover the sides of prothorax in many Orthoptera: in certain Hymenoptera, lie on each side of the parapsidal furrows of mesoscutum and = scapulae.

Lateral longitudinal area: of Hymenoptera, extends between the median and pleural carinae of metanotum.

Lateral ridge: in slug caterpillars, extends longitudinally along the lateral series of abdominal tubercles.

Lateral scale: one of the lateral processes of the ovipositor in Cynipidae, lying within and below the anal scale.

Lateral space: in slug caterpillars is the area on each side of the body between the subdorsal and lateral ridges.

Lateral sutures: of the thorax in Odonata, are situated on the sides of thorax, the first separating the metepisternum from the mesepimeron; the second separating the metepisternum from the metepimeron; the first more or less obsolete in the Anisoptera.

Lateral tubercle: lateral on thoracic and abdominal segments of caterpillars: it is 3 of the abdomen, 2a of thorax: constant (Dyar).

Latericeous: = lateritius: q.v.

Lateritius: yellowish-red; yellowish brick color [pale clay yellow with a little red].

Laterodorsal: the point of junction of dorsum and pleurum.

Lateropharyngeal: applied to the 4th pair of salivary glands in bees; situated on each side of the pharynx.

Laterostigmatal: situated on the side, immediately above the spiracle.

Lateroventral: the point of junction of sternum and pleurum.

Latero-ventral metathoracic carina: in Odonata; forms the dividing line between the metepimera and the metasternum.

Latescent: becoming obscure or hidden.

Latreille's segment: the first abdominal segment of those Hymenoptera in which it is fused with the thorax:= median segment, propodeon, propodeum.

Latticed: = cancellate; q.v.

Latus: the side: broad.

Latuscula: the facets of the compound eye.

Leathery: having the appearance or texture of leather.

Lectotype: a co-type chosen, subsequently to the original description, to take the place which in other cases a holotype occupies.

Leg -s: the jointed appendages attached to the thoracic segments, used in walking: the organs of locomotion other than wings: unjointed organs of locomotion are pro-legs or false legs; q.v.

Legion: a group of genera, subequal to a tribe.

Legnum: the margin of a squama.

Lemniscate: ribbon-like: in the form of an 8.

Lenticular: round, doubly convex; like a lens or lentil.

Lepidoptera: scale-winged: an order of insects with spirally coiled haustellate mouth structures; head free; thorax agglutinate; transformations complete four scale-covered wings.

Lepidopteric acid: a green pigment obtained from the wing scales of Lepidoptera; a derivative of uric acid: see Lepidotic acid.

Lepidopteron: a butterfly or moth: one of the Lepidoptera.

Lepidotic: set with minute scales.

Lepidotic acid: a yellow pigment obtained from certain butterfly scales a derivative of uric acid: see Lepidopteric acid.

Lepis: a scale.

Leprous: with loose, irregular scales.

Leptiform: = compodeiform; q.v.

Leptos: small, fine.

Lethargic: torpid or inactive.

Leucine: a white crystalline compound, the product of animal decomposition, found in the malpighian tubes: as a color, cheesy white.

Leucocytes: pale, unicellular bodies, numerous in the insect blood.

Levator: a muscle that raises an organ or a part.

Levigate -us: with a smooth, somewhat shiny surface.

Liber: free.

Ligament: a band or sheet of tough, fibrous tissue between two parts or segments.

Ligneous -eus: wood brown [Vandyke brown].

Lignivorous: feeding upon wood or woody tissues.

Ligula: the central sclerite of the labium, borne upon the mentum, usually single, sometimes paired: often used as synonymous with "glossa" and "tongue": corresponds to the united laciniae of right and left maxillae: see also elytral ligula.

Ligulate: strap-shaped; linear, much longer than broad.

Lilacinous: lilac-colored [lilac].

Limaciform: having the form of a Limax or slug; said of larvae.

Limb: the circumference: the area surrounding the disc.

Limbate: when a disc is surrounded by a margin of different color.

Limbus: the area along the outer and posterior margin of wing beyond the closed cells; Homoptera, Cicada.

Limpid: clear and transparent: applied to wings and ornamentation.

Line: a narrow streak or stripe: as a term of measurement, one- twelfth of an inch; commonly used by English and early American authors.

Linea: a line or narrow stripe.

Linear: straight; in the form of a right line.

Lineate: marked with lines or streaks: lined.

Lineolet: a delicate fine line.

Lingua: the tongue; applied in Hymenoptera, to the ligula: in Lepidoptera and Diptera, to maxillary structures: has also been used for the hypopharynx, and that use might be adopted: a median organ of the hypopharynx in Apterygota.

Lingua spiralis: the spiral tongue of Lepidoptera: see glossa.

Linguiform: tongue-shaped: linear, with the extremities obtusely rounded.

Lingula: in Aleurodidae, a more or less slender tongue or strap-shaped organ, attached cephalad within the vasiform orifice: a term proposed by Leuckart for the ligula of the bees.

Lipochromus: without color.

Lipoptera: = Mallophaga; q.v.

Literate: ornamented with characters like letters.

Littoral: living along the sea-coast or in the shore debris: strictly, between tide marks.

Littoralia: Heteroptera that live in marshes.

Litura: an indistinct spot, paler at its margin.

Livid: yellowish gray with a violet tinge: greenish gray.

Lobate -us: divided by deep, undulating and successive incisions.

Lobe: any prominent rounded process or excrescence on a margin: specifically, the rounded, tooth-like processes on the margin of the pygidium of the Diaspinae: also applied to lateral expansions of the abdominal segments.

Lobes: of the maxilla; see galea (outer) and lacinia (inner): of the mentum in Coleoptera, are the lateral expansions shielding the base of the central organs.

Lobes of pronotum: in Orthoptera; the spaces or areas formed by three transverse impressions on the pronotum: that which borders the head is the anterior lobe, the hindmost is the posterior lobe, those intervening are the middle lobes.

Lobiform: shaped like a lobe or rounded process.

Lobulate: divided into, or with many small lobes or lobules.

Lobule: in Coccidae, one of the two distinct parts of which a lobe is sometimes composed.

Lobulus: the partly separated portion of the wings of some flies and of secondaries in some Hymenoptera: also used as = alula; q.v.

Lobus: of maxilla = galea; q.v.

Locomotion: organs of, are legs and wings.

Longicorn -ia: having the antennae as long or longer than the body; specifically the Cerambycid beetles.

Longitudinal: in the direction of the long axis.

Longitudinal veins: are those that extend lengthwise through the wing either directly from base or as branches of one that does start there: they are named or numbered, and differently in the different orders.

Loop: applied to that structure at base of innerside of primaries into which the frenulum of male moths is fitted: see retinaculum.

Looper: applied to geometrid and other caterpillars in which some or all the middle abdominal legs are wanting and which move by bringing tail to thorax and forming a loop of the intervening segments.

Lora: the chitinous bands connecting the submentum with the cardo of maxilla (Comst.): the submentum: small cords upon which the base of the proboscis is seated (Say): the anterior part of the genae at the edge of the mouth: the corneous processes to which the muscles flexing the mouth in certain Diptera are attached, and in that sense the palpifer of the maxilla: in Homoptera, the small sclerite at side of clypeus and front, extending laterally to the genae.

Lorum: in bees: the angular piece upon which the sub-mentum rests.

Lower austral zone: occupies southern part of United States from Chesapeake Bay to the great interior valley of California. Is interrupted by the continental divide in eastern Arizona and west New Mexico and divided according to conditions of humidity into an eastern or Austroriparian and western or lower Sonoran area.

Lower field: in termini; see costal field.

Lower fronto-orbital bristles: in Diptera: are on the lower part of front, above the antennae, along the orbit.

Lower margin: of tegmina (Thomas), is the costal or anterior margin of other authors.

Lower radial vein: in Lepidoptera (Holland) media 2 (Comst.).

Lower sector of triangle: in Odonata - = cubitus 2 (Comst.).

Lower Sonoran faunal area: comprises the most arid deserts of North America, beginning west of lat. 98 degrees in Texas: sends narrow arms into southern New Mexico, is interrupted by the Continental Divide; covers a large part of w. and s. Ariz., s. w. Nev., s. w. Calif., a portion of central Calif., and most of Lower Calif. These areas are irregular and incapable of brief definition.

Lubricate -ous: covered with a slippery mucus.

Lucid: shining; applied to luminous insects.

Luciferase: a substance in the nature of an enzyme, existing in the luminous organs of light-giving beetles.

Luciferine: a substance in the blood of luminous beetles which, when brought into contact with luciferase, produces light.

Luciferous: light giving.

Lucifugous: fleeing the light: applied to nocturnal forms or those that live in concealment.

Lumen: the cavity of an organ: the inner surface of a tube: the hollow portion of a gland or vesicular structure.

Luminescence: applied to the light of fire-flies, as a substitute for phosphorescence.

Lumper: one who, in describing species or genera recognizes only prominent or obvious characters to the exclusion of minor color or variable characters of maculation or structure: see splitter.

Lunaris or Lunate: crescent-shaped: formed like a new moon.

Lunula: a small lunule or crescent.

Lunulae: in Hymenoptera, crescent-shaped marks near the orbits.

Lunulate: a line, when made up of a series of small lunules.

Lunule: a lunate mark or crescent.

Lurid -us: dirty brown with a bluish tinge [pale brown + a little French blue]: also used to indicate an obscuring of bright colors.

Luteo -testaceous: dark clay yellow.

Luteous -eus: clay yellow [pale clay yellow].

Lutescent: becoming or appearing to be clay yellow.

Lutose -us: apparently or really covered with dirt.

Lymphatic: producing, carrying or relating to the lymph.

Lyrate: lyre-shaped: cut into several transverse segments, and gradually enlarging towards the extremity.

Lyre: the upper wall or border of the spinning tube of caterpillars.

M

Macrochaetae: the long bristles occurring singly on the body of Diptera.

Macropterous: long or large winged.

Macrosomites: the primitive regions of primitive hand of the insect embryo.

Macula: a colored mark larger than a spot; of indeterminate figure.

Maculate -ed: spotted or marked with figures of any shape, of a color different from the ground.

Maculation: the ornamentation or pattern of marking.

Maculose: spotted; with many marks or spots.

Maerianum: "that segment of the post-pectus situate one on each side behind the acetabulum and parapleurum; it supports the posterior feet": see meriaeum.

Magenta: pinkish red; an aniline product.

Magis: more.

Maggot: applied to the footless larvae of Diptera.

Mala: a lobe: a ridged or grinding surface.

Mala mandibularis: the grinding surface or area of a mandible.

Mala maxillae: the globes of maxilla; outer or galea, inner or lacinia; where only one is present, the term refers to that one.

Malaxation: a kneading or softening; applied to the chewing and squeezing by fossorial wasps of insects captured as food for their larva.

Male: that sex having organs for the production of spermatozoa: designated by "?", the astronomical sign for Mars.

Mallophaga: wool-eaters: an ordinal term applied to biting lice: wingless: mandibulate; thoracic segments similar; no metamorphosis: =Lipoptera.

Malpighian tubules: long, slender tubules, varying in number, serving as excretory organs, entering the alimentary canal at the point of junction of chylific ventricle and ileum: said to be analogous with kidneys: = biliary vessels.

Mammilate: with nipple-like protuberances or processes.

Mandible: the lateral upper jaws of a biting insect.

Mandibular strobe: a broad deep groove on outer side of mandible in some Coleoptera.

Mandibular segment: the fourth or mandible bearing segment of head.

Mandibulata: that series of insects in which the adults have functional mandibles used for biting.

Mandibulate: with jaws or mandibles.

Manicate -us: fur-like: surface clothed with irregular depressed hair.

Manitrunk: that part of trunk that bears the anterior legs: =prothorax.

Manometabola: with a slight or gradual metamorphosis and without a resting stage; e.g. the Orthoptera.

Manubrium: in Coleoptera: that part of the mesosternum in Elateridae which forms the process for fitting into the cavity of the prothorax: in Collembola the basal part of the furculum.

Manus: the hand: formerly applied to the anterior tarsus.

Marbled: irregularly mottled, gray and white, like marble; = marmoratus.

Marcescent: shrivelling.

Margaritaceous: shining, like mother of pearl = nacreous; q.v.

Margin: that portion of a surface within the edge, bounded on the inner side by the sub-margin and consisting of a more or less dilated imaginary line.

Marginal: of, belonging to, or near the margin.

Marginal area: in Orthoptera; see mediastinal area.

Marginal bristles: in Diptera; are inserted on the posterior margin of the abdominal segment.

Marginal cell: in Diptera (Williston):= subcostal (Shiner):= radial (Comst.): in Hymenoptera:= radial and 2 (Comst.): in general that cell beyond the stigma.

Marginal field: in tegmina = costal field: q.v.

Marginal nervure or vein: in Orthoptera, = costa (Comst.): in Hymenoptera (Norton) = radius 3 (Comst.): in general, the vein forming the marginal cell.

Margined -ated: bounded by an elevated or attenuated margin: when the margin is edged by a flat border.

Marmorate -us: spots and lines irregularly disposed, as in marble: marbled. Mask: in the nymphs of Odonata, the modified labium which, when at rest, conceals the other mouth parts.

Masticate: to chew.

Masticatory: formed for chewing or grinding; applied to the mouth parts and to the grinding structures in the gizzard.

Mastigium -ia: telescopic anal organs in certain caterpillars, serving to repel attacks of parasites.

Matrix: the formative substance from which cells and other structures are derived.

Maxilla: without any qualifying adjective, the second pair of jaws in a mandibulate insect; the most persistent when the mouth is modified, and represented by some functional part in all insects in which the mouth structures are useful: second maxillae, = the labium, or third pair of jaws in a mandibulate insect.

Maxillary: attached or belonging to the maxilla; e.g. palpi.

Maxillary palpi: the first pair of palpi, borne on the maxilla.

Maxillary pleurites: the lateral pieces, epimera and episterna of the maxillary segment.

Maxillary segment: the sixth segment of the head, bearing the maxillae.

Maxillary tendons: two slender rods in basal third of the muscid proboscis the remnant of the palpifer, to which muscles for flexing the proboscis are attached: see lora.

Maxillary tentacle: in female Pronuba: a specialized process of palpifer.

Maxillulae: a pair of appendages in Thysanurids, between mandibles and first maxillae.

Maxime: very much or very large.

Mealy: with a flour-like dusting: = farinose.

Mecaptera: = Mecoptera, q.v.

Meconium: the substance excreted by certain metabolic insects soon after their emergence from the chrysalis or pupa.

Mecoptera: long-winged: neuropterous insects with similar, large, unfolded wings; mouth mandibulate, prolonged into a beak: head free; thorax agglutinated; transformations complete: the scorpion flies or Panorpidae. Medi-: prefix, = middle.

Media: the fourth of the longitudinal veins extending from base through approximately the middle of the wing, not more than four branched, the branches numbered on margin from nearest apex, to 4 nearest anal angle: in Orthoptera; it is the median or externomedian: in Lepidoptera (Pack.), is cubitus (Comst.).

Mediad: toward the median plane or middle.

Mediafurca: a process extending internally from the meso-sternum, to which the muscles are attached.

Medial: referring to, or at the middle.

Medial cells: (Comst.), are anteriorly bounded by the media or its branches: in Hymenoptera (Mort.), includes median and cubital (Comst.)

Medial cross-vein: (Comst.), is between media 2 and 3.

Median 1: in Lepidoptera (Pack.), = media 2 (Comst.).

Median 2: in Lepidoptera (Pack.), = media 3 (Comst.).

Median 4: in Lepidoptera (Pack.), = cubitus 2 (Comst.).

Median area: of wings in Orthoptera, lies between the radial and ulnar veins, radius and media (Comst.): of meta-thorax of Hymenoptera, is the middle of the dorsum, divided into three spaces or cells; 1st or basal area, 2d or Lipper median or areola; 3d or apical or petiolar area.

Median carina: Orthoptera; of head, is usually applied to a median dorsal carina, but has been also used for that which extends down the middle of front from the fastigium, and then = frontal costa: of prothorax, extends along the middle of pronotum.

Median cell: in Lepidoptera, is the closed area formed by a line extending from the end of subcostal to the end of the median veins, = radial (Comst.): in Hymenoptera, 1st median (Pack.), = medial (Comst.); 2d median (Pack.), - medial 4 (Comst.); 3d median (Pack.), = medial 2 (Comst.); 4th median (Pack.), = medial 1 (Comst.).

Median cross-veins: in Odonata; are those which cross median space.

Median foveola: in Orthoptera; the foveate depression of the vertex between the eyes: = central foveola.

Median forks: in Orthoptera, refers to the forks of the median vein.

Median furrow: lies between radius and media: in some Heteroptera, separates the embolium from the remainder of the corium.

Median lines: on the primaries of many moths: the first or t.a. crosses about one-third from base; the second or t.p. crosses beyond the outer third and is usually sinuate.

Median lobe: of labium in Odonata, is the partly divided glossa or ligula; probably corresponds to united glossa and paraglossae (Butler).

Median longitudinal carinae: on the metanotum of Hymenoptera, extend one on each side of the middle.

Median nervules: in Lepidoptera (Holland)1st = cubitus 2 (Comst.): 2d cubitus 1 (Comst.); 3d = media 3 (Comst.).

Median notch: in Coccidae, a notch in the edge of the pygidium, at the posterior extremity of the body.

Median plate: in Hymenoptera := sessiliventres, is the dorsal plate connecting the thorax and abdomen.

Median sector: in Odonata, = media 3 (Comst.).

Median segment: applied to the basal segment of the abdomen when it forms part of the metathorax: see propodeum.

Median shade or line: in Lepidoptera, crosses at or about middle of wings.

Median space: in Lepidoptera, is the area between the median lines: in Odonata, the cubital cell (Comst.); the space at base between submedian (radius) and postcosta (st anal); by Selys in 1896 and later used in the sense of medial cell of Comst.

Median vein: in Odonata and Lepidoptera, = radius (Comst.): in Lepidoptera, it runs from base to about middle, nearly through centre, and is four or five branched: in Hymenoptera, it is the 3d from costal margin.

Mediastinal: relating to the longitudinal median line or area.

Mediastinal area: in Orthoptera, the area between median or mediastinal vein and the costal or front margin: = marginal area.

Mediastinal vein: in Orthoptera and Diptera, = suhcosta (Comst.): also, in Diptera, = auxiliary vein (Meigen).

Medio-eubital cross-vein: between media 4 and cubitus, connecting the two series (Comst.).

Medio-ventral line: in caterpillars, extends along middle of under side.

Medipectus: the under side of meta-thorax: the mesosternum.

Mediproboscis: the middle third of the flexed proboscis of muscid flies.

Medi-thorax: =mesothorax; q.v.

Medius: middle.

Mega- Megalo-: large.

Melanic: with a blackish suffusion.

Melanism: an abnormal or unusual darkening: a suffusion with blackish.

Mellifera: honey-makers: applied to bees as a whole.

Melliferous: honey-producing, or producers of honey.

Mellisugous: honey-sucking: a feeder on honey.

Member: any one of the external appendages.

Membranaceous: thin, skin-like, semi-transparent, like parchment: of a thin, pliable texture.

Membrana retinens: the stretched part of the membrane around the rectum of butterfly larvae, used in the change to the chrysalis.

Membrane -ana: any thin, transparent, flexible body tissue: specifically the wing tissue between the veins: in Heteroptera, the thin membranous tip of the hemelytra.

Membranous or eous: composed of membrane or skin-like tissue.

Membranule: the small opaque expansion at base of wings in Odonata.

Meniscoidal: with one side concave the other convex, like a round segment from a hollow sphere.

Menognatha: insects in which both young and adults feed by mandibles; e.g. the Orthoptera: see menorhyncha and metagnatha.

Menorhyncha: forms in which both young and adult take food by suction e.g. Hemiptera: see metognatha and metagnatha.

Mental suture: in Coleoptera, the line between mentum and gula.

Mentigerous: bearing or having a mentum.

Mentum: a labial sclerite bearing the movable parts; attached to and sometimes fused with the sub-mentum; corresponds to the (united) stipes of maxillae: in Coleoptera, what is usually called mentum is really submentum: in Diptera, the term is applied to the posterior oral margin: in Hymenoptera, is part of "tongue," the second joint bearing the labial palpi, paraglossae and ligula.

Merdivorous: feeding upon dung or excrement: see scatophagous.

Meriaeum: the posterior inflected part of the metasternum in Coleoptera.

Meroistic: ovaries that secrete yolk or vitellaginous cells as well as ova.

Mesad: extending or directed toward the median plane.

Mesal: pertaining to, situated on or in the median plane of the body.

Mesenchym: that portion of the mesoderm that produces the connective tissues of the body.

Mesenteron: the mid-gut, stomach or chylific ventricle: the middle portion of the primitive intestinal canal, lined with entoderm.

Mesepimeron: in Odonata: the sclerite between humeral and first lateral suture.

Mesepisterna: in Odonata,- the oblique lateral pieces of mesothorax, meeting dorsally in a ridge.

Mesially: at or to the middle.

Mesinfraepisternum: a sclerite formed between propleuron, mesepisternum, mesepimeron and second coxa.

Meso: middle: as prefix, drops the o when stem begins with a vowel.

Mesoblast: the middles germ layer of the embryos: = mesoderm.

Mesoderm:= mesoblast: gives rise to muscular and circulatory systems.

Mesodont::= amphiodont: q.v.

Mesomeros: the 2d to 5th abdominal segments in Lepidoptera.

Meson: the middle plane of the body.

Mesonotum: the primitively upper surface of the 2d or middle thoracic ring.

Mesophragma: an internal prolongation of the metapraescutum, affording attachment to some of the wing muscles.

Mesopleura: in Diptera, the space before the root of the wing between the dorso- and sternopleural sutures: in Hymenoptera, the piece below the insertion of the wings.

Mesopleural bristles: in Diptera, are inserted in the angle formed by the dorsopleural and meso-pleural sutures.

Mesopleural suture: in Diptera, runs from the root of the wings downward and separates the meso-pleura from the pteropleura.

Mesopleuron: the lateral surface of the meta-thorax.

Mesosternal cavity: in Elateridae, the opening into which the prosternal spine or mucro is fitted.

Mesosternal epimera: in Coleoptera; the narrow pieces separating the metasternal from the meta-sternal episterna.

Mesosternal episterna: Coleoptera; on each side of mesosternum between anterior border and epimera; generally separated by a distinct suture.

Mesosternal lobes: in Orthoptera; = mesosternellum, q.v.

Mesosternellum: in Orthoptera, two median lobes of the mesosternum, one on each side of the deep median notch: in general, the sternellum of the mesothorax.

Mesosternum: the underside or breast of the meta-thorax.

Mesostethidium: = meso-thorax: q.v.

Mesostethium: the middle piece of the underside of meta-thorax, between the middle and hind legs.

Mesostigma: in Odonata, the spiracles of second thoracic segment.

Mesosulcus: a central longitudinal furrow of mesosternum in Hymenoptera.

Mesotarsus: the tarsus of the middle leg.

Mesothoracotheca: the pupal covering of the meso-thorax.

Mesothorax: the second or middle thoracic ring; bears the middle legs and the anterior wings.

Mesotergum: = mesonotum; q.v.

Meta-: posterior: used as a prefix to designate the third thoracic ring and its parts.

Metablastic: relating to the ecto- or meta-blast or ectoderm.

Metabola: insects with a complete metamorphosis in which the larva does not resemble the adult, and the pupa is quiescent.

Metabolism: is transformation: the whole process or series of changes of food into tissue and cell-substance and of these latter into waste products the first of these changes being anabolic, the second katabolic.

Metabolous: undergoing metamorphosis or transformation.

Metacoxal plate: in Coccinellidae, that portion of the first ventral segment included above the ventral lines visible on that segment.

Metagnatha: insects which feed with jaws when young and by suction, with tubular mouths when mature; e.g. the Lepidoptera: see menognatha and menorhyncha.

Metagonia: the hind or anal angle of a wing.

Metallic: having the appearance of metal: applied to a surface or color.

Metaloma: the sutural or inner margin of primaries.

Metamere: a segment, somite or athromere.

Metameric: made up of segments or metameres.

Metamerism: the arrangement in metameres.

Metameros: in Lepidoptera. the 6th to 8th abdominal segments.

Metamorphosis: is that series of changes through which an insect passes in its growth from egg through larva and pupa to adult: it is complete when the pupa is inactive and does not feed; incomplete when there is no pupa or when the pupa is active and feeds.

Metamorphosis dimidio: an incomplete transformation.

Metamorphosis perfecta: a complete transformation.

Metanotum: the primitively upper surface of the third or posterior thoracic ring: in Diptera, the oval arched portion behind, beneath the scutellum best developed in flies with long, slender abdomen: e.g. Tipulidae.

Metaphragma: the hindmost internal thoracic septum.

Metapleura: in Diptera, a swollen space at the outside of the metanotum, between it, the pteropleura and the hypopleura; in Hymenoptera, the piece behind and below the insertion of the hind wings.

Metapleural bristles: in Diptera, are inserted in the metapleura.

Metapneustic: larva, chiefly dipterous, in which the spiracles are confined to the posterior segment.

Metapnystega: that circular area of metanotum behind the postscutellum.

Metapodeon: the abdomen behind the podeon or petiole in Hymenoptera.

Metasternal: relating or attached to the metasternum.

Metasternal epimera: small sclerite separating the metasternal episterna from the ventral segments.

Metasternal episterna: sclerite situated on each side of the metasternum, immediately behind the mesosternum epimera.

Metasternellum: the sternellum of the metathorax.

Metasternum: the underside or breast of the metathorax.

Meta-stethidium: = meta-thorax; q.v.

Metastigma: in Odonata, the spiracles of third thoracic segment.

Metastoma: in Orthoptera:= hypopharynx: q.v.

Metatarsus: applied to basal joint of tarsus, where that differs greatly in length or otherwise from the other joints: see sarothrum.

Metatergum: = metanotum; q.v.

Metathoracotheca: the pupal covering of the meta-thorax.

Metathorax: the third thoracic ring or segment; bears the hind legs and second pair of wings; variably distinct; sometimes closely united with the mesothorax and sometimes appearing as a portion of the abdomen.

Metatype: is a specimen named by the author after comparison with the type; according to some, it should be also a topotype.

Metazona: in Orthoptera, the dorsal surface of the prothorax behind the principal sulcus.

Metepimeron: in Odonata, lies behind the second lateral suture and extends ventrally to the sternum.

Metepisternum: in Odonata, is the sclerite between the first and second lateral thoracic sutures.

Meter: the standard of length in the metric system = 39.37 inches: see centimeter and millimeter.

Meticulose -us: is a maculation in the form of a series of colored flames.

Metinfraepisternum: in Odonata; the sclerite just above base of 3d coxa; below metepisternum and before metepimeron.

Metochy: the relation borne to ants by the tolerated guests in ant- hills; demanding nothing from and giving nothing to the ants; see symphily and synecthry.

Metopidium: the anterior declivous surface of prothorax in Membracidae.

Micans: shining: also a surface of which only parts are shining.

Microchaetae: small bristles, as opposed to macrochaetae, in Diptera.

Microergates: the dwarf workers among ants.

Micron: the unit of microscopic measurement = 001 mm.: represented by the symbol μ: the symbol μμ represents .001 of a micron. {*Scanner's note: the μμ notation would no longer be valid.*}

Micropterous: small winged.

Micropterism: the tendency to produce small wings; applied to a line of variation.

Micropyles: minute openings in the egg, through which spermatozoa enter.

Microsomites: small secondary rings or somites of the macrosomites in the embryo, which afterward become the body segments.

Microthorax: a supposed thoracic ring between the head and prothorax.

Middle apical area: = internal area; q.v.

Middle field: = discoidal field; q.v.

Middle lobes: of pronotum in Orthoptera; see lobes.

Middle pleural area: in Hymenoptera; the median of the three areas between lateral and pleural carinae: = 2d pleural area.

Mid-dorsal thoracic Carina: a ridge or elevated line at the meeting of the mesepisterna in Odonata.

Mid-gut: the chylific ventricle with the caecal glands, tubes or pouches.

Mid-intestine: = mid-gut.

Migrants: applied to that brood of plant lice which flies from one to an alternate food plant: any forms that fly from the place where they were born for food or other purposes.

MM.: = Millimeter: .001 meter = .039 of an inch: roughly 25 mm. are counted to an inch in measuring insects. {*Scanner's comment: modern usage is lower case. So: mm.*}

Mimetic: when a species mimics or resembles another or some other object in appearance; but not in structure and other characters.

Mimicry: strictly, the resemblance of one animal to another not closely related animal, living in the same locality; often loosely used to denote also resemblance to plants and inanimate objects: Batesian mimicry is where one of two similar species is distasteful (so-called model), the other not distasteful (so-called mimic); Müllerian mimicry is where both species are distasteful.

Mines: applied to galleries or burrows between upper and under surface of leaf tissue, when made by larvae: they are linear, when they are narrow and only a little winding; serpentine, when they are curved or coiled, becoming gradually larger to a head-like end: trumpet- mines, when they start small and enlarge rapidly at tip; blotch mines, when they are irregular blotches tentiform, when the blotch mines throw the leaf into a fold on one side.

Miniate -us: of the color of red lead [vermilion with a slight admixture of dragon's blood].

Mirror: in Cicada; see specular membrane.

Mitosoma: the middle piece of a developing spermatozoon.

Mobile: movable: having the power of motion.

Model: see mimicry.

Modioliform: globular, truncated at both ends; like the hub of a wheel.

Mola or Molar: the ridged or roughened grinding surface of the mandible: when the mandible is compound, the molar corresponds to the subgalea of maxilla.

Monarsenous: that kind of union where one male suffices for many females.

Moniliform: beaded like a necklace.

Monochromatic: of one color throughout.

Monodactyle: with a single movable claw which closes on the tip of the other leg structures as in some parasitica.

Monodomous: ants in which each colony has one nest only.

Monoecious: when both sexual elements or glands exist in one individual.

Monogamous: a union where a female is fertilized by one male only.

Monomeri: insects with one-jointed tarsi.

Monomorphic: species of which only one sex (female) is known to exist.

Monophagous: insects feeding upon only one species or genus of plants.

Monothelious: a union where one female is fecundated by many males.

Monotrocha -ous: Hymenoptera in which the trochanters are single: having legs in which the trochanter is one-jointed.

Monotypical: a genus described from a single species, no other being known; or described from a single specified species with which are associated others believed to be identical in structure: see isotypical and heterotypical.

Moult: a period in the transformation when the larva changes from one instar to another: the cast skin of a larva that has moulted.

Mouth: the anterior opening into the alimentary canal, where the feeding structures are situated and in which the food is prepared for ingestion.

Mouth-parts: a collective name including labrum, mandibles, maxillae, labium and appendages = trophi.

Mucoreus: mouldy: a surface covered with small, fringe-like processes.

Mucro: a long, straight or curved process terminating in a point: the pro-sternal process in Elateridae: the terminal spine or process of an obtect pupa: "the median posterior point of the epigastrium when differentiated by elevation."

Mucronate: terminated in a sharp point.

Mucrones: in Collembola the two small end pieces of the furcula, proceeding from the dentes.

Mullerian association: a group of species belonging to different genera, often different families or even orders, having similar colors, possessing more or less distasteful qualities and living in the same locality.

Muller's thread: the common terminal thread of all the ovarian tubes.

Multangulate: with many angles.

Multi-: many; used as a prefix, often without the i.

Multiarticulate: with many joints or segments.

Multilocular: with many large cells, spaces or cavities.

Multipartite: divided into many parts.

Multiplicate: with many longitudinal folds or lines of plication.

Multispinose: with many spines.

Mumia: the pupa.

Munite -us: armed; provided with an armature.

Muricate -us: armed with sharp, rigid points.

Murinus: mouse colored [gray with some yellow].

Mushroom bodies: two stalked, mushroom-like bodies arising from procerebral lobes; supposed to be the seat of insect intelligence.

Muscle: the fleshy fibres of the insect body that serve to move the appendages and other body organs.

Mute: silent: without power to produce audible sound.

Mutic -us: unarmed: lacking processes where such usually occur.

Mutici: Acridiids without a posternal spines.

Mutilate -us: cut off: mutilated: abbreviated: not complete.

Mycetophagous: feeding upon fungi.

Myiasis: disease or injury caused by the attack of dipterous larvae.

Myoblast: a cell that produces muscular tissue.

Myrmecology: that branch of entomology that deals with ants.

Myrmecophilous: ant-loving: applied to insects that live in ant nests.

Mystacine -us: bcarded: with a hairy fringe above mouth or on clypeus.

Mystax: in Diptera; a patch of hair or bristles above the mouth, on the lower part of the hypostoma above the vibrissae.

Mytiliform: shell-like; as the middle feet in some aquatic Hemiptera.

N

Nacreous: pearly: resembling mother of pearl := margaritaceous.

Nail: a tarsal claw: specifically the stout pointed claws in predatory Heteroptera = unguis.

Naked: not clothed: lacking vestiture: a pupa when not inclosed in a cocoon or other covering.

Nasal suture: =clypeal suture; q.v.

Nasus: anterior termination of the face in certain Hymenoptera: the clypeus or a modification of it: in Odonata, the upper portion of the clypeus = supra-clypeus = postclypeus.

Nasuti: that type of termite soldiers that have the head prolonged into a point.

Natatorial -ions: formed for swimming.

Navicular: boat-shaped = cymbiform.

Neanic: referring to the pupal stage.

Nearctic: temperate and arctic North America, including Greenland.

Nebula: a cloud: a vague, indefined, dusky shading.

Nebulous -ose: cloudy: without definite form or outline.

Neck: the slender connecting structure between head and thorax of such insects as have the head free: any contraction of the head at its juncture with the thorax.

Necrophagous: living in or on carrion.

Nectaries: honey-tubes, cornicles, siphuncles; q.v.

Nematid: thread-like.

Nematocera: = nemocera; q. A.

Nematocerous: with long, thread-like antenna.

Nemocera: Diptera with long, at least six-jointed antennae.

Nemoglossata: bees with a thread-like tongue.

Nemoricolous: living in open, sunny woods.

Neogeic: belonging to the Western Hemisphere or New World: see gerontogeic.

Neolepidoptera: all haustellate Lepidoptera, except the generalized Micropterygidae; mandibles not functionally present; pupa incomplete or obtect: see paleolepidoptera and protolepidoptera.

Neoteinic: applied to complemental females in Termites because, though reproductive, they retain some juvenile characters.

Neotropical: that part of the earth's surface embraced in the greater part of Mexico, West Indies and South America.

Neotype: a specimen identified with a species already described, and selected as a standard of reference where the original type or co-types are lost or destroyed.

Nephridia: tubular structures functioning as kidneys in Annelids, Mollusks, etc. and incorrectly used as = malpighian tubules; q.v.

Nepionic: that stage of development immediately succeeding the embryonic; proposed as a substitute for larval.

Nerinaeum: a ventral thoracic sclerite between the metasternum and posterior coxa in some Coleoptera.

Nerve: a thread-like structure, composed of delicate filaments whose function it is to transmit sensations or stimuli to or from a ganglion or from or to any part of the body or its appendages.

Nerves: sometimes used to = veins, in wing structures.

Nervi: belonging or referring to the nerves.

Nervulation: arrangement of the nerves: specifically applied to the arrangement of the chitinous framework of wings and thus= venation; q.v. Nervules or Nervures: the rod or vein-like structures supporting the membranes of wings and = veins and veinlets; q.v.

Nervuration: = nervulation and venation: q.v.

Neural canal: an incomplete tunnel on the floor of meso- and metathorax, formed by fusion of apodemes, serving for the reception and protection of the ventral nerve cord and for the attachment of muscles.

Neural groove: is that furrow in the primitive layer of the embryo in which the nerve cord is formed.

Neuration: = venation; q.v.

Neurilemma: the external sheath of a nerve fibre.

Neuroblast: the large cell in the early embryo, from which the nervous system develops.

Neuromere: that part of a body segment pertaining to the nervous system.

Neuroptera: nerve-winged: an ordinal term applied to insects with four net-veined wings; mouth mandibulate: head free: thorax loosely agglutinated; metamorphosis complete: in its older use, the term applied to all net-veined insects irrespective of metamorphosis or thoracic structure.

Neuropteroidea: like the Neuroptera in the wide sense; applied to those living insects included by Linnaeus in his Neuroptera; also to those extinct forms which have a general resemblance to them.

Neurospongium: a granular matrix in the periopticon of the insect eye.

Neuter: the term applied to workers or undeveloped females in some Hymenoptera.

Nidificate: to nest: applied when eggs are placed in a prepared receptaculum.

Niger: black.

Nigricans: black, tinged with gray.

Nits: the eggs of sucking lice; specifically when attached to a hair: in general, though rarely, applied in the singular to an egg.

Nitidus: shining: applied to a highly polished, smooth surface.

Niveous -eus: snowy white.

Nocturnal: species that fly or are active at night.

Nodal furrow: in Odonata; a transverse suture, beginning at a point in costal margin corresponding to the nodus, and extending toward inner margin.

Nodal sector: in Odonata; = media 2 (Comst.): arises from upper sector of arculus near nodus and extends to outer margin.

Node: a knot or knob: in the plural refers to the small segment or segment between thorax and main portion of abdomen in ants.

Nodiform: in the form of a knot or knob.

Nodicorn: with antennae that have the apex of each joint swollen.

Nodose -us: knotted or with knots; a body with one or more knotted parts a sculpture with almost isolated knots.

Nodule: a little knot, lump or node.

Nodulose -us -ate: with small nodes or nodules: a surface sculpture of knots or links, connected by an undulating line.

Nodus: in Odonata; a stout, oblique, short vein at the place where the anterior margin of the wings is sometimes drawn in.

Nopalry: a plantation of cacti for raising cochineal insects.

Normal: of the usual form or type: not out of the ordinary.

Notate: marked by spots: with a series of depressed marks as a sculpture.

Notched: indented, cut or nicked; usually a margin.

Notocephalon: in some aquatic Hemiptera, that part of the head which is apparent from a dorsal aspect.

Notodont: with toothed backs: applied to a series of moths whose larvae are more or less conspicuously humped on dorsal surface.

Notopleural suture: = dorso-pleural suture; q.v.

Nototheca: that part of the pupa covering upper surface of abdomen.

Notum: the dorsal or upper part of a segment: = tergum.

Nucha: the upper surface of the neck connecting head and thorax.

Nucleate: with, or having a nucleus.

Nucleolus: the small portion of matter in the nucleus most readily affected by staining fluids.

Nucleus: a well-defined, differentiated, round or oval body imbedded in the cell contents.

Nude -us: naked: a surface devoid of hair, scales or other vestiture.

Nuditas: = nudity.

Nudity: the state of being naked or bare of vestiture.

Nurses: worker ants or worker bees which care for the eggs, larvae and pupae, but do not forage, the latter function being taken up later, when nursing is given up.

Nutant: nodding; the tip bent toward the horizon.

Nutritive chamber: an enlarged section of ovarian tube, filled with granular nutritive material used in developing the egg cells.

Nymph: the larval stage of insects with incomplete metamorphosis: applies also to their pupal stage, and sometimes used as = pupa.

Nympha inclusa: = coarctate pupa; q.v.

Nymphipara: applied to insects that bear living young in an advanced stage of development: see also pupipara.

O

Ob-: as a prefix, means inversely.

Obconic: conic, with the apex pointing downward.

Obcordate: inversely heart-shaped, with the point applied to the base of another object or part.

Obese -us: unnaturally distended: usually applied to the abdomen.

Oblate: flattened; applied to a spheroid of which the diameter is shortened at two opposite ends.

Oblique: any direction between perpendicular and horizontal.

Oblique vein: in Odonata; an apparent cross-vein situated between M2 and Rs, distal to the level of the nodus and inclined obliquely, from its front end, backward and outward; in reality the basal part of Rs.

Obliterate: nearly washed out; indistinct.

Oblong: longer than broad.

Obovate: inversely egg-shaped; the narrow end downward.

Obpyriform: inversely pearshaped.

Obscure: not readily seen: not well defined.

Obsite-us: a surface covered with equal scales or other bodies.

Obsolete: nearly or entirely lost: inconspicuous.

Obtect: wrapped in a hard covering.

Obtected: applied to pupae when they are covered with a chitinous case which confines and conceals all appendages, though their outlines may be marked on the surface: see free, and coarctate.

Obtuse: not pointed: an angle greater than a right angle: opposed to acute. Obtuse-angulate: two markings or margins meeting so as to form an obtuse angle.

Obtusilingues: short-tongued bees with the tip obtuse or bifid: see acutilingues.

Occipital foramen: the opening in the occiput, opposed to a similar opening in the prothorax: = foramen magnum.

Occipital margin: in Mallophaga, the posterior margin of the head.

Occipito-orbital bristles: in Diptera; situated on posterior orbit of eye.

Occiput: that part of the head behind the vertex: in Diptera, the whole posterior surface of the head: in bees, the space between the vertex and the neck.

Occlusor: applied to muscles which close an opening; e.g. spiracles.

Occult -us: hidden; concealed from superficial view.

Ocellar bristles: in Diptera, are situated close to the ocelli, usually directed forward: often absent.

Ocellar ribband: a crescent-shaped, smooth thin belt across the eye region in butterfly chrysalids.

Ocellar triangle: a triangle, indicated by grooves or depressions, on which the ocelli are situated; Diptera.

Ocellate: eye-like in appearance: in Lepidoptera, spots on the wings, bordered by a colored iris or ring, and usually with a pupil.

Ocelli: plural of ocellus; q.v.; = stemmata.

Ocelligerous: supplied with, or bearing ocelli.

Ocellus: a simple eye, consisting of a single convex or bead-like lens, which conveys an image to a retina.

Ocelli occur in larvae and, singly or in small groups, in adults: the compound eyes are made up of numerous ocelli.

Ochraceous: yellow with a slight tinge of brown [pale cadmium yellow and brown ochre].

Ochraeus -eus: = ochraceous.

Ochro-leucus: dilute ochraceous.

Ocular emargination: in Mallophaga, a lateral emargination of the head in which the eye is received posteriorly.

Ocular fleck: in Mallophaga, a small, intensely black spot of pigment in the eyes.

Ocular fringe: in Mallophaga, closely set small hair on posterior half of ocular emargination, sometimes extending on temporal margin.

Ocular lobes: of brain = procerebrum; q.v.

Ocular sclerite: the first or protocerebral segment of the head.

Ocular tubercles: in Aphids, are a group of prominent facets on the hinder part of each eye.

Oculi -us: the eyes: an eye: refers to the compound eyes.

Oculocephalic: applied to that pair of imaginal buds destined to produce the cephalic region in Hymenoptera.

Odona: toothed: applied to Odonata by Fabricius because of the long teeth on the maxilla and labium.

Odonata: net-veined insects with mandibulate mouth; head free; thorax agglutinate; wings similar, elongate, flat; metamorphosis incomplete; copulatory organs of male near base of abdomen, separate from the testes.

Odonate: bearing toothed mouth parts, like those of dragon flies.

Odoriferous: diffusing an odor; applied to glands or secreting organs.

OEcology: see ecology.

OEdagus: the penis.

OEnocytes: large yellow cells arranged segmentally in clusters, in each side of body cavity: associated with blood and fat bodies.

OEsophageal bone: a plate below anterior part of oesophagus in Psocidae.

OEsophageal bulb: = sub-clypeal pump; q.v.

OEsophageal diverticula: = food reservoirs (q.v.); but more generally applied also to any sac-like structure connected with the gullet.

OEsophageal lobes: form posterior portion of brain or tritocerebrum.

Oesophageal valve: a funnel-like folding of the oesophagus, extending into the chylific ventricle in some insects, and forming a valve that controls the entrance of food into that organ: = cardiac valvule.

Oesophagus: the gullet: that part of the alimentary canal between the mouth and the crop.

Olfactory: pertaining to the sense of smell: those lobes of the deutocerebrum from which the nerves supplying the antennae arise.

Oligonephria: applied to insects with few urinary (Malpighian) tubes.

Oligoneura: having few wing veins: specifically applied in Diptera to Cecidomyids.

Olivaceous: with a tinge of olive-green, usually as a shading [olive green].

Omaloptera: the pupiparous flies.

Omia: the shoulders: the lateral anterior angles of an agglutinated thorax, when they are distinct:= see umbone: in Coleoptera; a corneous sclerite to which the muscles of the anterior coxa are attached; also the lateral margin of the prothorax; also the lateral margin of the scutellum in Carabids and Dytiscids.

Ommateum: the compound eye.

Ommatidium -ia: one of the elements of which the compound eye is composed.

Omnivorous: a general feeder upon animal or vegetable food, or both.

Oncus -i: a welt: applied to welt-like ridges on caterpillars.

Onisciform: shaped like a wood-louse, Oniscus sp.; applied to certain Lycaenid and other caterpillars.

Ontogenetic: relating to the development of the individual.

Ontogeny: the development of the individual as distinguished from that of the species: see phylogeny.

Onyches: claws of tarsi.

Onychium -ia: small processes between the tarsal claws in many Diptera; see empodium: a more or less retractile process on the feet of some beetles: in Hymenoptera, the apical tarsal joint bearing the claws: see also arolium and pulvillus.

Oöblast: the primitive germinal nucleus of an egg.

Oögenesis: the process of egg-formation.

Oölemma: the cell wall of an egg: see vitelline membrane.

Oötheca: the covering or case over an egg mass, as in certain Orthoptera: see egg case.

Opacus: opaque; a surface without any lustre.

Opalescent: with a bluish white lustre, as in opals.

Opalinus: = opalescent; q.v.

Opaque: without lustre: not transparent.

Operaria: the workers in Hymenoptera.

Operative: in working order or actually working.

Opercula: two plates covering the vocal structure of Cicada, beneath.

Operculum: a lid or covering: in Diptera, the chitinous envelope covering the lower part of the muscid mouth; the labrum-epipharynx of Dimmock: the scutes covering the meso-thoracic stigmata: in Aleurodidae, the lid-like structure closing the vasiform orifice; q.v.

Ophthalmic: relating to the eye.

Ophthalmotheca: that part of the pupa that covers the eyes.

Opisthogoneate: having the organs of generation at hind end of body.

Opisthogonia: the anal angle of the secondaries.

Opposite: placed over against, or opposed to.

Optic: relating to the organs of vision.

Optic ganglia: are at the sides of the procerebrum and innervate the compound eyes.

Optic lobes: the laterals lobes of the procerebrum in which are centered the nerves supplying the organs of vision.

Opticon: the first of a series of three ganglionic swellings in the optic nerve: see epiopticon and periopticon.

Optic segment: =procerebral segment; q.

Optic tract: is the perceptive portion of the compound eye.

Ora: a border: specifically in some Coleoptera, the lateral margin of prothorax.

Ora coleopterorum: the margin of the elytra.

Orad: toward the mouth.

Oral: pertaining to the mouth.

Oral cavity: the mouth; = buccal cavity.

Oral fossa: in Mallophaga, a furrow lying in front of the mandibles.

Oral segment: that ring or segment which bears the mouth.

Orbicular: round and flat, the diameters of the plane equal: in sonic moths, a round or oval macula in the median cell.

Orbit: an imaginary border around the eye: in Diptera the orbits are divided into vertical or superior; frontal and facial or anterior; of the cheek or inferior; occipital or posterior.

Orbital sclerite: a narrow sclerite encircling some eyes.

Order: one of the primary divisions of the Class Insecta, based largely on wing structure and then usually ending in -ptera.

Ordure: excrement; usually applied to such as is foul or offensive.

Orichalceous: = aurichalceous; q.v.

Oriental: in geographical zoology as used by Wallace, that part of the earth's surface including Asia east of the Indus River, south of the Himalayas and the Yangtse-kiang watershed, Ceylon, Sumatra, Java and the Philippines.

Orificium: the anal or genital opening.

Original type: is the actual specimen from which a published description is prepared.

Orismologia -y: the defining of scientific or technical terms.

Orthoptera: straight winged: an ordinal term applied to insects in which the primaries are not used in flight, but cover the longitudinally folded secondaries; mouth mandibulate; head set into prothorax, the latter free; metamorphosis incomplete.

Orthorrhapha: that section of Diptera in which the pupa escapes from larval skin through a T-shaped opening on back: see cyclorrhapha.

Orthorrhaphous: straight-seamed.

Os: the mouth of insects, in general.

Oscillation: a vibrating or swinging from side to side.

Osculant: intermediate in character between two groups or series.

Osmeterium -ia: fleshy, tubular, eversible processes producing a penetrating odor, capable of being projected through a slit in the prothoracic segment of certain Papilionid caterpillars, and from openings elsewhere in the bodies of other forms.

Osmosis: the tendency of liquids to pass or diffuse through a membrane or septum.

Osselet: = ossicle; q.v.

Ossicle: a small nodule of chitin resembling a bone.

Ossicula: small corneous pieces that serve in the articulation of the wings to the thorax.

Ostia: the slit-like openings of the heart.

Ostiolar canal: a marginal furrow leading from the ostiole.

Ostiole: in Heteroptera, the openings at the sides of meso- and meta- thorax, through which an odoriferous fluid is excreted.

Ostium: singular of Ostia; q.v.

-osus; an affix, signifying saturation, or the possession of the quality expressed in the stem word.

Otocyst: an auditory or ear-like vesicle.

Otolith: a little ear-bone: granules or concretions found in an otocyst.

Outer lobe: of maxilla = galea; q.v.

Outer margin: the outer edge of wing, between apex and hind angle.

Ovo, Ovum: the eggs; an egg.

Ova glebata: eggs laid or concealed in lumps of dung.

Ova imposita: eggs laid in the substance that is to serve as food for the larva.

Oval: egg-shaped, with both ends similar.

Ova pilosa: eggs that are covered with hair: usually from the abdomen of the female.

Ovarian tube: a tubular structure in which are developed the cells forming the future ova: a single one of the mass which, taken together, form the ovaries.

Ovaries: a mass of ovarian tubes, lying one on each side of the body cavity of the female, in each of which tubes eggs or ova are developed: the individual tubes of an ovary all converge to one oviduct.

Ovariole: an ovarian tube: q.v.

Ovary: singular of ovaries; q.v.

Ovate: in outline, egg-shaped or oval.

Oviduct: the tube through which the egg passes from ovarian tubes into vagina: sometimes used in the sense of ovipositor: q.v.

Oviform: egg-shaped.

Oviparous: where reproduction is through eggs laid by the female.

Oviposition: the act of depositing the eggs.

Ovipositor: the tubular or valved structure by means of which the eggs are placed; usually concealed; but sometimes extended far beyond the end of the body.

Oviscapt: = ovipositor; q.v.

Ovivalvule: in Ephemeroptera; is an appendage of the female reproductive organs.

Ovoviviparous: when living young are born from eggs which are hatched in the body of the parent.

P

Pad: the pulvillus, or that part of it which is capable of extension and retraction in some Coleoptera.

Paddle: the flattened joints of posterior tarsi in aquatic Hemiptera.

Paedogenesis: reproduction in the sexually immature or larval stage.

Paedogenetic: reproducing in the sexually immature or larval stage.

Pagina: the surface of a wing: P. superior, is the upper surface; P. inferior, the lower surface: in Orthoptera, the external flattened surface of the caudal femora.

Pagiopoda: Heteroptera, in which the posterior coxae are not globose and the articulation is a hinge joint: see trochalopoda.

Pagiopodous: those Heteroptera which have the coxae of the hind legs hinged and the femora grooved.

Pala: the shovel-shaped tarsal joints in many aquatic Heteroptera.

Palate: = hypopharynx; q.v.

Paleace: chaff or chaffy: = paleaceous.

Paleaceous: chaffy in appearance.

Palearctic: relating to that part of the earth's surface including Europe, Africa north of Sahara, and Asia as far south as the southern edge of the Yang-tse-Kiang watershed and the Himalayas, and west to the Indus River.

Paleodictyoptera: an ordinal name suggested by Scudder for Paleozoic insects which cannot be assigned to existing orders.

Paleolepidoptera: haustellate Lepidoptera in which the mandibles are distinct and the pupa is free: includes the Micropterygidae only: see protolepidoptera and neolepidoptera.

Pallescent: becoming pale or light in color or tint.

Pallette: the disc-like structure composed of three tarsal joints, on the anterior feet of male Dytiscidae.

Pallid: pale or very pale.

Pallide-flavens: pale or whitish yellow.

Pallidus: of a pale, cadaverous hue [a very dilute brown pink].

Pallium: an erectile membrane partially closing the open cavity formed by the walls of the sub-genital plate in Melanopli.

Palma: the basal segment of the anterior tarsus when it is broadened or specifically modified.

Palmate: like the palm of the hand, with finger-like processes.

Palmula: = pulvillus; q.v.

Palp: a mouth feeler or palpus.

Palpal: belonging, relating or attached to the palpi.

Palparium: in some Coleoptera, and other insects, the membranous support to which the labial palpi are attached, and which permits an amount of extension not possible when they are fixed.

Palpi: plural of palpus; q.v.

Palpicorne: with long, slender, antenna-like palpi.

Palpifer: any palpus-bearing part: specifically, a small sclerite hearing the maxillary palpus and itself articulated to the stipes.

Palpiferous or -gerous: bearing a palpus.

Palpiger: that sclerite of the labium to which the labial palpus is attached corresponds to the palpifer of the maxilla and has been used in the same general sense.

Palpigerous stipes: in Coleopterous larvae, = palpifer; q.v.

Palpuli: the maxillary palpi in Lepidoptera, when visibly developed.

Palpus: a mouth feeler: tactile, usually jointed structures borne by the maxillae (maxillary palpi) and labium (labial palpi).

Panduriform: violin shaped: oblong, with rounded ends, medially constricted.

Panorpatae: = Mecoptera; q.v. Pantherine: in color, almost like cervinus; q.v.: in maculation, like those of a panther. Papilioform: formed like a butterfly wing.

Papilionaceous: butterfly-like.

Papilla: a minute, soft projection: specifically the modified ligula in silk spinning caterpillars.

Papillary: with nipple-like processes that have the tips rounded.

Papillate -us: a surface with small elevations which are porous at tip.

Papilliform: like a wart or pimple.

Papillose -us: pimply; a surface covered with raised dots or pimples.

Pappose: downy: made up or clothed with pappus.

Pappas: a fine down.

Para-: next to; near by; at the side of.

Parabiosis: see symbiosis.

Parabolic: elongately rounded.

Paraclypeal piece: in lepidopterous pupa, occurs in some of the generalized families on each side of the maxillary palpi.

Paraclypeus: in caterpillars, a narrow sclerite bordering clypeus at sides.

Paraderm: the limiting membrane enclosing the pronymph of Muscidae.

Paraglossa: a paired, labial structure, lying at each side of the ligula; often connected with it; sometimes free and two-jointed: corresponds to the galea of maxilla.

Parallel: along the same line and nearly equidistant.

Paranal: at the side of or next to the anus or anal structures.

Paranal forks: two lateral, bristle-like structures in some caterpillars, used to throw frass pellets to a distance.

Paranal lobes: = podical plates; q.v.

Paraphysis: the chitinized thickenings or lateral ingrowths, usually situated at the base of the lobes in certain Diaspid genera.

Parapleura: the sternal side pieces in beetles.

Parapodia: the pro- or false legs: more specifically applied to the jointed abdominal processes of the Symphyla.

Parapsidae: the small sclerites on each side of the scutellum in Chalcids, marked by the parapsidal grooves.

Parapsidal furrows: longitudinal grooves on each side of the mesoscutum of Proctytrypidae separating the parapsides from the middle lobe.

Parapsidal grooves: the grooves or furrows on each side of the Chalcid scutellum, defining the parapsidae.

Parapsides: lateral pieces of the meso-scutum, separated from the mesal portion by the parapsidal furrows.

Parapteron -era: small sclerites, articulated to the dorsal extremity of the episternum, just below the wings; absent on prothorax = the tegulae of Hymenoptera, and patagia of Lepidoptera: have been homologized with the elytra of Coleoptera.

Parasita: = parasitica: q.v.

Parasite: a species that lives in or on another animal or insect, and depends upon the tissue of the host for its food supply.

Parasitic: living on or in some other animal or insect in such a way as to derive all nourishment from the tissues of the host.

Parasitica: the sucking lice: wingless; without metamorphosis; mouth with piercing lancets; thoracic segments similar; habits epizoötic.

Parasitism: a form of symbiosis in which one party lives upon or at the expense of the other, makes no return and destroys its host: see symbiosis; commensalism.

Parastigma: = pterostigma; q.v.

Parastigmatic glands: small, circular glands, which secrete a waxy powder, sometimes present around the spiracles of Coccidae.

Paratype: is every specimen of the series from which the type was selected see type and cotype.

Parse: sparse or sparsely.

Parcidentate: with few teeth.

Parenchymatous: composed of soft cellular and connective tissue.

Parietes: walls: the perpendicular sides of elevated bodies.

Paronychium-ia: one or more bristle-like appendages of onychia; q.v.

Parthenogenetic: see asexual.

Parthenogenesis: reproduction by direct growth of germs from egg-cells without fertilization by the male element: as in plant lice, gall wasps, etc.

Particolored: partly of one, partly of another color: divided into two or more color fields.

Partite -us: divided; e.g. the eyes of Gyrinidae. Parum: not much.

Patagium -ia: in Lepidoptera, those sclerites that cover the base of primaries: often used as synonymous with tegula and squamula, q.v.: assigned by some writers to the pro-, by others to the meso-thorax: homologized with the paraptera of meso-thorax.

Patella -ae: the modified joints of anterior tarsi in Dytiscidae; plate- like, horny or spongy structures on the undersides of the tarsal joints: the first coxal joint.

Patellar: pertaining to the knee-joint or cap.

Patellariae: in Dytiscids, the unequal, cup-like impressions on the underside of the patella.

Patens, Patentes: open; diverging; spreading apart.

Patric: home or country of origin.

Patulous-ose: open, spreading.

Paunch: a crop-like accessory pouch in some Mallophaga: any pouch- like appendage of the alimentary canal.

Paurometabolous: metamorphosis in which the changes of form are gradual and inconspicuous: e.g. Orthoptera and most Rhynchota.

Pavillions: the sheds or cells sometimes built by ants as a shelter for groups of plant lice.

Paxilla: a small stake or peg: a bundle of spicular processes.

Pearlaceous: having the appearance of pearl.

Pecten: a comb: in Hymenoptera, rigid, incurred setae on the basal parts of maxilla and labium: the rows of spines on the feet of pollen- gathering bees: any series of bristles arranged like a comb: in mosquito larvae the comb-like teeth on the breathing tube.

Pectinate: comb-shaped: with even branches like the teeth of a comb.

Pectinato-fimbriate: having pectinations that are fringed with hair.

Pectoralis: relating to the breast.

Pectoral plate: in Coleoptera, the sternum.

Pectunculate: with a row of minute appendages like the teeth of a comb: e.g. some maxillary structures.

Pectus: the ventral portion of thorax: variably applied in Coleoptera, for the entire meso- and meta-thorax: also the pro- and meso-sternum: in Diptera, is the inferior surface of the thorax between the legs.

Pedal line: in caterpillars: extends along the base of the feet.

Pedal tubercle: on the thoracic and abdominal rings of caterpillars: on the anterior side of leg-base and, correspondingly, on apodal segments: is VII of the abdomen where it consists of three setae: VI of the thorax where the setae are not numbered: constant (Dyar).

Pedamina: the aborted fore-legs of Nymphalid butterflies.

Pedate: foot-bearing, or having feet.

Peddler: applied to the larvae of such Cassid beetles as carry their excrement and cast skins on an anal fork.

Pedes: the feet, or really, legs.

Pedicellus or Pedicle: the third joint in a geniculate antenna: forming the pivot between scape and funicle: in general, a stalk or stem.

Pediculosis: a state of lousiness, or the abnormal condition caused by the multiplication of lice on the body: sec phthiriasis.

Pediculous: lousy: infested with lice.

Pedigerous: feet bearing.

Peduncle: a stalk or petiole: the basal joint of the antenna in Homoptera: the smaller of the two stalks supporting the mushroom body; q.v. Pedunculated: set on a stalk or peduncle: attached by a slender stalk or neck.

Pelagic: inhabiting the sea, far from land.

Pellicles: the exuviae or cast larval skins of many insects: in Coccidae more especially applied to the hardened larval skin attached to the puparia of Diaspinae.

Pellit: covered with long, drooping hairs, irregularly placed.

Pellucid: colored, but transparent: sometimes applied when there is no color.

Pelotons: the balls of fine tracheae in larvae, developed to supply the adult organism.

Pelottae: =arolia: q.v.

Peltate: shield- or target-shaped.

Penal claspers: in Proctytripidae. lateral fringed processes of the male genitalia.

Penal sheath: the horny outer covering of the penis.

Pencil: a little, elongated brush of hair: in Diptera, applied to a group of sensory hairs on the flagellum of the antenna.

Pendent: hanging down.

Pendulous: drooping: hanging free, attached to one end only.

Penes: open, slit-like structures of the seminal vesicles to the outer surface in Euplectoptera.

Penicillate: with a long, flexible brush or pencil of hair: often at the end of a thin stalk.

Penicilli: a pair of small style or cerci-like pieces on the tip of the 8th dorsal segment of abdomen of various male Hymenoptera.

Penicilliform: pencil-like or shaped.

Penicillum: a pencil or brush of long hair attached at the end of a stalk as long as the brush, and folded in a lateral groove in some male moths.

Penis: the flexible, membranous, intromittent organ of the male.

Pennaceous: = pennate.

Fermate: feathered or bearing feather-like processes.

Penniform: feather-like in form.

Pentagon -um: a five-sided figure with five equal or unequal angles.

Pentamera: Coleoptera with 5-jointed tarsi. Pentamerous: species having five-jointed tarsi.

Penultimate: next to the last.

Peptone: a soluble proteid compound produced by the digestion of album-menoid food substances.

Per-: as a prefix, means very: extremely: through.

Percipient: with the power of perceiving.

Percurrent: running through the entire length.

Pereion: the prothorax.

Pereipoda: the second and third pair of thoracic legs of larvae, and the 2d pair in adults.

Perfoliate: divided into leaf-like plates: applied to antennae with disc-like expansions connected by a stalk passing nearly through their centres: also to any part possessing a well-developed leaf-like or plate-like expansion.

Pergamenous: thin, partly transparent: resembling parchment.

Peri-: round about. Periopticon: a complex nerve structure back of the basilar membrane of the eye.

Pericardial: around, or belonging to the heart.

Pericardial cavity: the space between the diaphragm and dorsal body wall, which contains the heart.

Pericardial cells: specialized cells, which lie along both sides of the heart, and whose function it is to purify the blood.

Pericardial chamber: is the open space around the heart or dorsal vessel.

Pericardial diaphragm: a delicate membranous tissue attached to the ventral surface of the heart and laterally to the body wall := dorsal diaphragm wings of the heart.

Peri-intestinal: that part of the body cavity around the alimentary canal.

Peri-neural: situated around a nerve: the body cavity immediately surrounding the nervous system.

Periodical: recurring at regular intervals.

Periopticon: third ganglionic swelling of optic tract: see opticon.

Peripheral: referring to the outer margin.

Peripheria: the entire outline of the body.

Periphery: the circumference or outer margin.

Peripneustic: larvae which have the spiracles absent on middle and posterior thoracic rings, and present on all other body segments.

Peripodal cavities: pouches in the embryo in which the rudiments of the future legs and wings are developed.

Peripodal membrane: the cell layer surrounding the peripodal cavities. Peristaltic: that periodic motion of the alimentary canal by means of which the food is forced toward the anal extremity.

Peristethium: the meso-sternum.

Peristoma -ium: the border of the mouth or oral margin in Diptera; sometimes used as := epistoma: q.v.

Peristome: a membranous tissue surrounding the mouth parts at base, and forming the true ventral wall of the head.

Peritoneal: applied to the membrane surrounding the viscera, trachea, and other internal structures.

Peritracheal: surrounding the trachea.

Peritreme: the corneous selerite surrounding a spiracle.

Peritrophic membrane: a funnel-like extension of the fore-gut, extending back tube-like, through the chylific ventricle in some insects.

Perivisceral: the cavity containing the alimentary canal and its appendages.

Perlate: beaded: bearing relieved, rounded points in series.

Perpendicular: upright: at right angles to horizontal.

Persicinus: the red of peach blossoms.

Persistent: remaining constantly; always present.

Personate: gaping wide open; masked; disguised.

Pes, Pedes: a foot feet.

Petiolar area or Petiolarea: on the metanotum of some Hymenoptera, the apical or hindmost of the three median cells 3d median area; apical area.

Petiolate: that series of Hymenoptera in which there is a slender stalk between the thorax and abdomen: = apocrita.

Petiolate: supported or placed on a stem or stalk; usually applied in describing venation and the method of attachment of abdomen to thorax.

Petiole: a stem or stalk: specifically the slender segment between the thorax and abdomen in many Hymenoptera, and some Diptera.

Phaeism: applied to a duskiness of butterflies occurring in a limited region.

Phagocyte: a corpuscle or cell that devours or absorbs noxious organisms and also absorbs the organs of the larval stage in the developments to the adult condition.

Phagocytosis: the destruction or devouring of bacteria or other microorganisms by phagocytes.

Phalaenae: a Linnean term embracing most of the heterocerous Lepidoptera: more specifically applied to the Geometridae.

Phalanx -ges: a joint or joints of the tarsus: a division of classification of uncertain value: similar to tribe.

Phalerated: beaded.

Phallus: =penis: q.v.

Pharyngeal pump: = sucking pump; q.v.

Pharynx: the back part of the mouth and upper part of the throat: a slight enlargement at the beginning of the oesophagus: in Diptera is sometimes restricted to the space between the hypopharynx and sub-clypeal pump, and is then = sub-clypeal tube.

Phauloptera: an ordinal term for the scale insects (Laporte 1835).

Phleboptera: = Hymenoptera; q.v.

Phonetic: sound producing; applied to stridulating structures.

Phosphorescent: shining or glowing in the dark, like phosphorus.

Photogenic: a light producing structure; producing a Phosphorescent glow.

Photoscopic: characters of light or conditions of illumination that affect colors of caterpillars.

Phragma: a partition or dividing membrane: longitudinal, thin partitions passing down from the dorsum of meso- and meta-thorax: the partition formed by the inflexed hinder edge of prothorax.

Phragmocyttares: social wasps in which the combs of the nest are wholly or partly supported by the covering envelope: see stelocyttares: poecilocyttares.

Phthiriasis: a diseased condition of the skin caused by sucking lice.

Phyllophagous: feeding upon leaf tissue.

Phylogenetic: relating to tribal or stem development.

Phylogeny: the development of a genus, family, tribe or class: see ontogeny.

Phyloptera: the super-ordinal term proposed to include all the net-veined orders, the Orthoptera and Dermatoptera.

Phylum: a stem or tribe: used in classification to indicate a series of related organisms.

Physopoda: bladder-footed: = Thysanoptera; q.v.

Phytophaga: plant-eaters: beetles in which the 4th and 5th tarsal joints are anchylosed and the 3d is lobed.

Phytophagus: feeding upon plants.

Phytophilous: plant loving: species that live on plants.

Phytophthira: plant lice: some authors include also scale insects.

Piceous -eus: pitchy black.

Picine: black, with a bluish oily lustre.

Pick: a chitinous maxillary structure in Psocidae.

Pieza: the combined biting and sucking mouth of the Hymenoptera.

Piezata: the Fabrician term for Hymenoptera.

Pigment: any coloring matter or material that gives a color appearance.

Pile: a hairy or fur-like covering: in Diptera, applied to thick, fine, short, erect hair, giving a surface appearance like velvet.

Pilifer or Piliger: a small sclerite at each side of the clypeus in Lepidoptera, resembling a rudimentary mandible.

Piliferous: with a covering of fine hair or pile.

Pillared eye: in Ephemerids, that type which is placed on a cylindrical stalk or process: = turbinate eye.

Pilous or Pilose: clothed with down, or dense pile: with long, sparse hair.

Pilosity: a covering of fine, long hair.

Pincers: the anal forceps.

Pinna: a narrow wing; a feather.

Pinnae: of posterior femur in jumping Orthoptera, are the oblique ridges running to the median line and somewhat resembling a feather.

Pinnate: feather-like; cleft, like the wings of Alucita: with markings resembling a feather: with stiff hairs or thorny processes occupying opposite sides of a thin shank.

Pinnatifid: divided into feathers, as when wings are cleft nearly to the base.

Pistazinus: yellowish green, with a slight brownish tinge [pale green with a little burnt sienna].

Plaga: a spot, stripe or streak of color; a longitudinal spot of irregular form.

Plaited: longitudinally folded or laid in pleats.

Planate: with a flattened surface.

Plane: level, flat; applied to a surface.

Planipennia: applied to Neuroptera in which the wings are large and laid flat on the body wnen at rest; Sialidae, Myrmeleonidae, etc.

Planta -ae: the basal joint of the posterior tarsus in pollen gathering Hymenoptera: the soles of the posterior tarsal joints: the anal clasping legs of caterpillars.

Plantigrade: species that walk on the entire foot, not on the claws alone.

Plantula: a lobe of the divided tarsal pulvillus; one of the soles or climbing cushions of the foot: see arolium; pulviglus.

Plaques: the small leathery hemelytra in some Naucorids.

Plasma: the liquid portion of animal fluids and cells.

Plasticity: the capacity for being formed, moulded or developed.

Plate: any broad flattened piece or sclerite: = squame, in Coccidae.

Platelet: a little plate or sclerite of chitin in a membrane.

Plates: in Coccidae, the squames; q.v.: in male Homoptera, a pair of pieces following the last full ventral segment; usually preceded by a short piece, - the valve.

Platyptera: flat and broad-winged: an ordinal term applied to insects with four net-veined wings, secondaries longitudinally folded beneath primaries; mouth mandibulate; prothorax free; transformations complete.

Plecoptera or Plectoptera: plaited winged: an ordinal term applied to net-veined insects in which the secondaries are longitudinally folded beneath primaries; mouth mandibulate; body loosely jointed; prothorax free; metamorphosis incomplete.

Pleon: = abdomen; q.v.

Pleopoda: abdominal legs of larva: posterior legs of an adult.

Plesiobiosis: see symbiosis.

Plesiotype: any specimen identified with a described or named species by a person other than the describer.

Pleura: plural of pleuron or pleurum: the lateral sclerites between the dorsal and sternal portion of the thorax: in general, the sides of the body between the dorsum and sternum.

Pleural areas: on the metanotum of some Hymenoptera, the three spaces between the lateral and pleural carinae; the 1st or anterior = spiracular area; the 2d or central = middle pleural; the 3d or posterior = angular area.

Pleural carinae: in Hymenoptera, extend along the exterior margin of the metanotum.

Pleural pieces: the lateral sclerites of the thorax; see pleura.

Pleurites: the sclerites into which the pleurum is divided.

Pleuron: the side of the thorax.

Pleuropodia: embryonic or temporary bands formed by the modified first pair of abdominal legs in many insects.

Pleurostict: lamellicorn beetles in which the abdominal spiracles are situated on the dorsal portion of the ventral sclerites.

Pleurum: = pleuron; plural, pleura; q.v.

Plexus: a knot: applied to a knot-like mass of nerves, or tracheae.

Plica: a fold or wrinkle: a longitudinal plait of a wing.

Plicate: plaited; folded like a fan.

Plications: folding,; applied to the folds on the hind wings of Orthoptera.

Plicipenna: = Trichoptera; proposed by Latreille.

Plumate: like a feather.

Plumbeus: leaden or bluish gray [neutral].

Plumose: feathered; like a plume: antennae that have long ciliated processes on each side of each joint: see cirrate.

Plump: with full, rounded outlines; not obese.

Plumules: specialized scales of the androconia of male Lepidoptera.

Pluri: as a prefix, means many.

Pluri-dentate: with many teeth.

Pluri-setose: bearing several seta; as the head in some Carabids.

Pluri-valve: with several valves or valve-like appendages.

Pneumogastric: the ganglion supplying nerves for the tracheal and digestive system: also used as = vagus: q.v.

Pneustocera: breathing horns: the prolongations of the metathoracic spiracles in Berytidae, etc.

Pnystega: in Odonata, applied by Charpentier to a portion of mesonotum

Pobrachial: a longitudinal vein of the Ephemerid wing just behind praebrachial; usually simple: number 7 of some systems.

Podeon: in Hymenoptera, the petiole: the true second abdominal segment.

Podex: the upper plate of the anal opening; = supra-anal or sur-anal plate in caterpillars.

Podical plates: the latero-ventral plates attached to the loth abdominal segment of Orthoptera; the two pieces on each side of the vent; united they form the tergite of a rudimentary ring: = anal valves: paranal lobes.

Pododunera: apterous insects with biting mouth structures.

Podotheca: that part of pupa that covers the legs of future adult.

Poecilocyttares: social wasps that build their combs around the branch or other support covered by the envelope: see stelocyttares and phragmocyttares.

Poisers: = halteres and balancers; q.v.

Poison glands: sometimes applied to the salivary glands of bugs and biting flies; more usually to an abdominal gland connected with the sting of female Hymenoptera.

Policate: a tibia produced inwardly into a short, bent spine or thumb.

Politus: smooth, shiny, polished.

Pollen: a dusty or pruinose surface covering which is easily rubbed off; used mostly in Diptera.

Pollen-plate: a polished area margined by hair, on the outer face of the tibia in bees.

Pollex: a thumb: the stout fixed spur at inside of tip of tibia.

Pollicatus: = policate; q.v.

Polliniferous: formed for collecting pollen: pollen bearing.

Pollinigerous: = polliniferous: q.v.

Pollinose: covered with a yellow, pollen-like dust.

Poly-: many, much.

Polyandry: where a female mates with more than one male.

Polychromatic: many colored.

Polydomous: applied to ants when one colony has several nests.

Polyembryony: the production of several embryos from a single egg, as in some Chalcids.

Polygamy: where a male mates with more than one female.

Polygonal: with many angles.

Polygoneutism: the power to preduce several broods in one season.

Polymorpha: the claviform and serricorn Coleoptera, as a whole.

Polymorphicous: occurring in several forms; differing in sex, in season, in locality or without apparent reason: undergoing several changes, and in this sense applied to insects with a complete metamorphosis.

Polynephria: applied to insects with many urinary (Malpighian) tubes.

Polyphagous: eating many kinds of food.

Polyphyletic: derived or descended from several stems or sources.

Polypodous: having many feet, and thus, specifically applied to the Myriapoda, and to the larvae of Lepidoptera and saw-flies, in contradistinction to footless and hexapodous larvae.

Ponderable: that which may be weighed.

Pone: behind (the middle).

Ponticulus: = frenulum; q.v.

Porcate: marked with raised longitudinal lines.

Pore: any small, round opening on the surface.

Poriferous: closely set with deep pittings or punctures.

Porose -us: with little round openings on the surface.

Porrect: stretched out forward: straightly prominent.

Post-: behind or after.

Post-alar callosities: rounded processes at the posterior lateral margin of the dorsum, in Diptera.

Post-alar callus: in Diptera, a rounded swelling between the root of the wing and the scutellum.

Post-alar membrane: the strip of membrane connecting the squamae with the scutellum.

Postal vein: in Hymenoptera, = costa (Comst.).

Post-annellus: in Hymenoptera, the 4th joint of antenna and 2d of flagellum.

Post antennal organs: in Collembola, oblong or ellipsoidal organs situated just caudad of the bases of the antenna.

Post-brachial: = pobrachial; q.v.

Post-cerebral: applied to that pair of salivary glands in bees, situated close to the posterior wall of the head.

Post-clypeus: in Odonata, the upper of the two parts into which the clypeus is divided: in Psocidae, a peculiar inflated structure behind the clypeus: in general, the posterior or upper part of clypeus when any line of demarcation exists: = supra-clypeus; nasus: afternose; paraclypeus: first clypeus; clypeus posterior.

Post-costa: = subcosta (Comst.): in Odonata, = 1st anal vein (Comst.): in Trichoptera = anal.

Post-costal space: Odonata; the cell or cells lying posterior to the post- costa = anal cell (Comst.).

Post-cubitals: = post-nodal spaces; q.v.

Post-dorsulum: the middle piece of the meta-notum, between the mesophragma and post-scutellum.

Post-embryonic -otic: the stage after the insect has come out of the egg.

Post-epistoma: that part of the head behind the clypeus in Hymenoptera: see also post-clypeus.

Posterior: hinder or hindmost: opposed to anterior: in Diptera; applied to that face of the legs which is not visible when viewed from the front, the legs being laterally extended.

Posterior angle: of thorax, in Coleoptera, is the lateral angle near base of elytra: of the wings = hind angle; anal angle; q.v.

Posterior cells: in Diptera (Will.): 1st radial 5 (Comst.): 2d = medial 1 (Comst.): 3d = 2d medial 2 (Comst.) 4th = medial 3 (Comst.): 5th cubitus 1 (Comst.).

Posterior cephalic foramen: in Odonata, the opening of head posteriorly through which the cavities of head and thorax communicate.

Posterior field: of tegmina, = anal field; q.v.

Posterior intercalary: in Diptera, is one of the anal veins (Comst.).

Posterior lateral margins: in Orthoptera, extend from base of pronotum downward to the posterior angle of sides.

Posterior lobe: of the pronotum in Orthoptera, see lobe: in Diptera, that part of wing between axillary incision and base: = alar appendage (Loew). Posterior margin: = inner margin; q.v.

Posterior pereion: the meta-notum.

Posterior pleon: the terminal segments of the abdomen.

Posterior pleopoda: the anal clasping legs of caterpillars: see planta.

Posterior stigmatal tubercle: on thoracic and abdominal segments of caterpillars; varies in position from substigmatal to stigmatal posterior; sometimes united to V: it is IV of the abdomen, II of the thorax (Dyar).

Posterior trapezoidal tubercle: on the thoracic and abdominal segments of caterpillars; subdorsal, posterior, always present, rarely united with I: it is II of the abdomen, lb of the thorax (Dyar).

Posterior veins: those separating the posterior cells.

Posterior wings: = secondaries: q.v.

Postero-dorsal: Diptera; applied to leg bristles at the meeting of the dorsal and posterior face.

Postero-ventral: Diptera; applied to leg bristles at the meeting of the ventral and posterior face.

Postfurca: an internal process of metasternum to which the muscles of hind legs are attached.

Post-gena: the sclerite below occiput and behind gena in some Orthoptera.

Post-gula: is situated at the extreme base of the underside of the head in Dermaptera.

Post-humeral bristles: in Diptera, are usually two, inserted above the dorsopleural suture between the humeral callus and root of wing, on the bottom of the presutural depression.

Postical vein: in Diptera, = 5th longitudinal (Meig.); = media 3 (Comst.).

Posticus: hinder.

Post-media: Ephemerida; an apparently distinct vein between media and Cubitus (Comst.).

Post-medial line: in Lepidoptera, = t.p. line: q.v.

Post-median: Diptera; those leg bristles situated above or behind the middle. Post-nodal cross-veins: in Odonata, the transverse veins between costa and radius 1, and radius 1 and media 1, from nodus to stigma, separating the post-nodal cells or spaces: = post-cubital cross-veins.

Post-nodal costal spaces: in Odonata, the cells below costal margin from nodus to stigma.

Post-nodal radial spaces: in Odonata, the cells between radius 1 and media 1, from nodus to outer margin.

Post-nodal sector: in Odonata, a longitudinal vein lying between media 1 and media 2 (Comst.): = ultra-nodal sector.

Post-oral: behind the mouth; those segments bearing mouth structures.

Post-pectus: the under surface of the meta-thorax.

Post-petiole: in Hymenoptera, that part of abdomen behind petiole. Post-retinal: the fibres arising from the facets of the compound eye and extending into the ganglionic plate.

Post-scutellum: the fourth and posterior sclerite of the dorsum of the thoracic rings.

Post-sutural: in Trichoptera, the little plate behind the scutellum of mesothorax: = post-scutellum.

Poststigmatal: that portion of the marginal cell beyond the stigma in bees: = 2d radial 1 (Comst.).

Post-stigmatal primary tubercle: on thoracic segment of caterpillars; sub-primary, stigmatal, posterior; it is III of the thorax and not present on abdomen (Dyar).

Post-subterminal: following the s. t. line in Lepidoptera.

Post-sutural bristles: in Diptera, dorsal bristles behind transverse suture.

Post-terga: applied to the posterior scutes of the segments of Coleopterous larvae.

Post-triangular cells: = discoidal areolets; q.v.

Post-vertical cephalic bristles: in Diptera, are in the middle of upper part or Occiput.

Pouch: in Trichoptera, a depressed, usually longitudinal area in a wing.

Prae- or Pre-: anterior to; before.

Praebrachial: a longitudinal vein in middle of an Ephemerid wing; usually forked: no. 6 of some systems.

Praecostal spur: a false vein in costal angle at base of secondaries.

Prae-dorsum: = prophragma: q.v.

Prae-labrum: in Diptera = clypeus: q.v.

Praeocular: before the eyes.

Praeputium: the external membranous covering of penis: specifically a spherical muscular mass at base of penis in some Orthoptera.

Prae-scutellum: a sclerite, rarely present, between the meso-scutum and meso-scutellum.

Prae-scutum: the first of the four divisions of the notum of the thoracic rings.

Prae-subterminal: preceding the s.t. line in Lepidoptera. Prae-terga: the anterior thoracic scutes in coleopterous larvae.

Prae-tornal: preceding the tornus (q.v.) in Lepidoptera.

Prasinus: grass-green [apple green].

Pratinicolous: frequenting or living in grassy meadows or bogs.

Pre-alar callus: a small swelling or projection before the root of wings, just back of outer ends of transverse suture, in Diptera.

Pre-anal: above or before the anal opening.

Pre-anal plate or lamina: = supra-anal plate; q.v.

Pre-antennal: anterior to or before the antenna.

Pre-apical: before the apex.

Pre-balancer: = pre-halter: q.v.

Pre-basilar: before the base.

Precocious stages: generally applied to all stages of development from the fertilized egg to the pupa.

Precurrent: continuous: entire: complete: said of a vein.

Predaceous: applied to insects that live by preying upon other organisms.

Predatory: = predaceous: q.v.

Pre-eruciform: before the caterpillar stage: specifically applied to the early larvae of some Proctytrypidae.

Preformation: the doctrine of growth or development from already existing rudiments; opposed to epigenesis: q.v.

Pre-furca: "the stem vein in front of a fork, that reaches back to where itself forks from another vein"; Diptera.

Pregenicular: in Orthoptera, that portion of femur proximad the knee.

Pregenicular annulus: a more or less conspicuous color ring on the caudal femora proximad the knee in Orthoptera.

Pre-halter: a membranous scale in front of the true haltere of a fly.

Prehension: structures fitted for grasping or holding.

Pre-mandibular: situated in front of the mandible: applied to a temporary segment of the embryo: = intercalary segment.

Pre-media: Ephemeridae; an apparently distinct vein between radius and media (Comst.).

Premorse: as if bitten off: with a blunt or jagged termination.

Prensor: the genital lateral clasping organ of male Lepidoptera: see clasper.

Pre-ocular: see prae-ocular.

Pre-oral: in front of the mouth: the embryonic head segments before those bearing the mouth parts.

Prepuce: =praeputium; q.v.

Pre-pupal: that stage in the larva just preceding the change to pupa.

Pre-scutellar bristles: in Diptera, are in a transverse row in front of the scutellum.

Pre-scutellar callus: = post-alar callus: q.v.

Pre-scutellar rows: in Diptera, short rows of small bristles in front of the scutellum.

Press: =filator; q.v.

Pressure plate: a structure at base of pulvillus, which exerts a pressure on the sole of the pad.

Pre-sutural bristles: in Diptera, in a trigonate depression at outer ends of transverse suture, near dorso-pleural suture.

Pre-sutural inter-alar bristle: the single bristle of the interalar series, situated before the transverse suture.

Primaries: the anterior or fore-wings.

Primitive: simple in character; of an early or ancient type.

Principal sector: in Odonata, extends from its point of separation From the median sector to the outer margin, at or just below the apex:= media 1 (Comst.).

Principal sulcus: in Orthoptera, a transverse impression of the prothorax, at or behind the middle.

Priodont: applied to those forms of male Lucanids that have the smallest mandibles: see teleodont, mesodont, amphiodont.

Prismatic: formed like a prism: a play of colors similar to that produced through a prism.

Pro-: anterior: used as a prefix to designate the parts of the first thoracic segment.

Proboscidea: an ordinal term for the Coccidae.

Proboscis: generally applied to any extended mouth structure; usually applied to the extensile mouth of the Diptera; frequently to the beak of Hemiptera;

sometimes to the tongue of Lepidoptera; and rarely, to the mouth of long-tongued bees.

Procephalic: relating or belonging to the procephalon.

Procephalic lobes: in the embryo, form part of the anterior, overhanging portion of the head.

Procephalon: that segment of the head in the embryo which is formed by the coalescence of the first three primitive segments.

Procerebral: that segment of the brain containing the median protocerebrum and optic ganglia; also called optic segment.

Procerebral lobes: the central portion of the cerebrum, made up of the fused median lobes, giving rise to the mushroom bodies; q.v.

Procerebrum: the anterior part of the brain, formed by the ganglion of the first primary segment; also termed ocular lobe, froth part it innervates. Process: a prolongation of the surface, margin, or an appendage: any prominent portion of the body not otherwise definable.

Process of labrum: in bees = appendicle: q.v.

Procidentia: the narrow projecting tip of 7th dorsal segment in Nematinae.

Proclinate: directed forward; applied to hair or bristles.

Proctodaeum: the invagination of epiblast that produces the anus and intestine as far forward as and including malpighian tubes.

Proculiform or Poculiform: hollow, cylindrical, with a hemispherical base, the sides at top straight goblet-shaped.

Procumbent: trailing; prostrate; lying flat.

Produced: drawn out; prolonged; extended from.

Proeminent: said of the head when it is horizontal and does not form an angle with the thorax.

Profile: the outline as seen from the side.

Profound, Profundus: deep.

Prognathus: having the jaws directed forward.

Progoneate: with the genital opening on an anterior body segment.

Progonia: the anterior angle of the secondaries.

Proleg: any process or appendage that serves the purpose of a leg: specifically the fleshy unjointed abdominal legs of caterpillars and certain saw-fly larvae: = abdominal feet: false legs.

Proloma: the anterior margin of the secondaries.

Prolonged: extended or lengthened beyond ordinary limits.

Promeros: the first abdominal segment in Lepidoptera.

Prominent: raised or produced beyond the level or margin: standing out in relief by color or otherwise: conspicuous.

Promuscidate: with proboscis or extended mouth structure.

Promuscis: an extended mouth structure: has been applied to the long tongue of bees and to the rostrate structure in Hemiptera.

Pronotal carina: in Orthoptera, the main or median carina on pronotum.

Pronotum: the upper or dorsal surface of the prothorax.

Pronucleus: the nucleus of male and female elements, spermatozoa and ova, the union of which forms the nucleus of a fertilized ovum.

Pronymph: is that stage in certain inctabolous insects in which the larval tissues are completely broken down, and the imaginal tissues are just beginning to build up.

Proparaptera: the paraptera of the prothorax: the terms erroneously applied in this connection.

Propedes: = prolegs: q.v.

Prophragma: the anterior dividing wall of meso-thorax, which is horny and, at its upper edge, bears the connecting membrane between pro- and mesothorax.

Prop-leg: = pro-leg; q.v.

Propleura: the lateral portions of prothorax.

Propleural bristles: in Diptera, are situated immediately above the front coxa: = prothoracic bristle; q.v.

Propneustic: larval forms in which only the most anterior spiracles occur.

Propodeon: = propodeum.

Propodeum: in Hymenoptera, that part of thorax just above insertion of abdomen, and really the first abdominal segment: see median segment.

Propolis: a glue or resin-like product elaborated by bees to serve as a cement in cases where wax is not sufficiently tenacious.

Propulsatory: that which drives onward or forward.

Propupa: a semi-pupa: q.v.

Propygidium: the dorsal segment or tergite in front of the pygidilini, sometimes left exposed in Coleoptera.

Propygium: = Hypopygium: q.v.

Proscutum: the scutum of the pronotum.

Proscutellum: the scutellum of the pronotum.

Prostemmatic: = ante-ocular; q.v.

Prosternal: belonging to the prosternum.

Prosternal grooves: occur laterally in some Coleoptera. e.g. Elateridae, to receive the antennae.

Prosternal epimera: the epimera of prothorax.

Prosternal episterna: the episterna of prothorax.

Prosternal lobe: in some Coleoptera, an anterior prolongation of the prosternum which more or less conceals the mouth from below.

Prosternal spine: the curved mucro in Elateridae which extends backward into a meso-sternal cavity: the cone or tubercle between fore-legs in some Orthoptera.

Prosternal suture: that suture of pro-thorax which separates the sternum front the pleural pieces.

Prosternellum: the sternellum of the prothorax

Prosternum: the fore-breast: the sclerite between the fore-legs.

Prostheca: a mandibular sclerite set with hair, articulated to the basalis - q.v. -and equal to the lacinia of the maxilla.

Protamphibion: a name applied by P. Mayer to the hypothetical common ancestor of the Perlina, Ephemerina and Odonata.

Protandry: the appearance of males earlier in the season than females.

Protarsus: the tarsus of the anterior leg.

Proteiform: having many fortes or varieties: protean.

Protergum: in Odonata, the upper surface of prothorax.

Proterotypes: primary types, including all the material upon which the original description is based.

Prothoracic bristle: in Diptera, a strong bristle immediately above the front coxa: see propleural bristles.

Prothoracic glands: occur in Orthoptera, on the sides of prothorax in certain Phasmid genera.

Prothoracic shield: = cervical shield: q.v.

Prothoracotheca: the pupal covering of prothorax.

Pro-thorax: the first thoracic ring or segment: hears the anterior legs but no wings: when free, as in Coleoptera. is usually referred to as "thorax" merely.

Protocerebral segment: = ocular seginent; q.v.

Protocerebrum: the primitive anterior cerebral vesicle.

Protocosta: the thickened costal margin of Lepidopterous wings.

Protocranium: the posterior part of the epicranium: sometimes used as Occiput.

Protogonia: the apical angle of the primaries.

Protograph: all original description by a figure or picture made from the original type.

Proto-lepidoptera: proposed for those forms (Eriocephalidae) in which lacinia and mandibles are obvious and the spiral tongue is not developed: see neolepidoptera and paleolepidoptera.

Protolog: the original description by words.

Protoloma: the anterior margin of primaries.

Protomesal: applied to certain areolets in Hymenoptera, situated between costal cells and apical margin

Prototergite: the foremost dorsal segment of abdomen.

Preto-thorax: = prothorax: q.v.

Prototype: a primitive form to which later forms can be traced.

Protractor: that which extends or lengthens out: applied to muscles.

Protuberance: any elevation above the surface.

Protuberant: rising or produced above the surface or general level.

Proventriculus: the posterior portion of crop: the gizzard.

Proximad: toward the proximal end.

Proximal: that part of an appendage nearest the body: see distal.

Prozona: in Orthoptera, the upper or dorsal surface of prothorax in front of the principal sulcus.

Pruinose: hoary: as if covered with a fine frost or dust.

Pruinous -us: deep blue with a reddish tinge, like a plum [French blue + purple lake].

Psammophilous: living in sandy places.

Pselaphotheca: that part of the pupa which covers the palpi.

Pseudidolum: = nymph: q.v.

Pseudimago: = sub-imago; q.v. Pseud- or Pseudo-: as a prefix means false, spurious, or merely resembling. Pseudo-cellula: = accessory cell: q.v.

Pseudo-chrysalis: the semi-pupa.

Pseudo-coel: a false hollow; a hollow which does not form a tube.

Pseudo-cone: a soft, gelatinous cone in the compound eye of some insects, replacing the crystalline cone of others.

Pseudo-elytra: the aborted anterior wings of Strepsiptera.

Pseudogyna fundatrix: in Aphids, is the immediate issue of a fecundated egg: a stem-mother.

Pseudogyna gemmans: in Aphids. are wingless descendants of the stem-mother (fundatrix) or of the winged migrants (migrans) which reproduce asexually through a number of generations.

Pseudogyna migrans: in Aphids, the winged descendants of the stem- mother (fundatrix) through which the species is spread.

Pseudogyna pupifera: in Aphids, the last generation of p. gemmans, which produces the true sexes.

Pseudogyna: a female that reproduces without impregnation.

Pseudo-halteres: the rudimentary primaries of Stylops.

Pseudo-neurium: a false vein formed by a chitinous thickening of a wing fold.

Pseudo-neuroptera: those net-winged insects with incomplete metamorphosis: includes the present Ephemeroptera, Odonata, Plecoptera, Isoptera and Corrodentia: = Archiptera.

Pseudonychium -ia: = paronychia; q.v.

Pseudo-nymph: = semipupa; q.v.

Pseudopodia: = parapodia; q.v.

Pseudoptera: an ordinal name for the scale insects.

Pseudo-pupa: the inactive larval stage preceding the formation of the true pupa in some insccts; e.g. Meloidae: = semi-pupa; q.v.

Pseudo-pupillae: in Odonata, the black spots seen on the compound eyes of the living insects.

Pseudosessile: those petiolate Hymenoptera, in which the abdomen is so close to the thorax as to seem sessile.

Pseudo-trachea: the ringed and ridged grooves on the labella of Diptera, by means of which they scrape their food.

Pseudova: egg-like germ cells capable of development without fertilization e.g. in certain plant lice.

Pseudovary: the organ or mass of germ cells of an agamic insect.

Pseudo-vitellus: a cellular organ in Aphididae, supposed to replace the absent Malpighian tubules.

Psocoptera: = Corrodentia; q.v.

Psychogenesis: the origin and development of social and other instincts and habits. Pterodicera: with wings and two antenna.

Pterogostia: the wing veins.

Pterogostia: referring to the wing structure. Pteropega: wing sockets or cavities into which the wings are inserted.

Pteropleura: in Diptera, are situated below the base of the wings behind the meso-pleural suture: = the posterior lateral plate of meso- thorax of Lowne; the episternum of meso-thorax of Hammond.

Pteropleural bristles: in Diptera, are inserted on the pteropleura.

Pterostigma: a thickened, opaque spot on the costal margin of a wing, near its middle or at end of the radius: = bathmis, and see stigma.

Pterotheca: that part of the pupa that covers the wings.

Pterothorax: the wing-hearing thoracic segments in Thysanoptera.

Pterygium: a lateral expansion of the snout of some Coleoptera.

Pterygodes: the patagia or tegtila: q.v.

Pterygogenea: insects that are winged in the adult stage or believed to be descended from winged ancestors: see apterogogenea.

Pterygostium: a wing vein.

Pterygote: wing bearing.

Ptilinum: in Diptera cyclorrhapha, an inflatable organ capable of being thrust out through a frontal suture just above the root of antenna.

Ptilota: winged insects.

Pubes or Pubescence: short, fine, soft, erect hair or down.

Pubescent: downy: clothed with soft, short, fine, closely set hair.

Pubis: the lateral region of the prothorax.

Pulmonarium: the membranous connection of the plates or scutes of the abdominal rings: = connexivum.

Pulsatile: having the power of pulsating or moving in a rhythmic manner: applied to special organs in the legs, which aid in circulating the blood in these appendages.

Pulverulent: powdery or dusty in appearance.

Pulvilliform: having the appearance or structure of a pulvillus.

Pulvillus -i: soft, pad-like structures between tarsal claws: the cushions of short, stiff hair or other clothing on the underside of tarsal joints; rarely fleshy lobes: see arolium.

Pulvinatus: moderately convex.

Punctate: set with impressed points or punctures.

Punctiformis: shaped like a point or dot.

Punctulatus: with small punctures.

Puncture: an impression like that made by a needle.

Punctured: marked with small, impressed dots.

Puniceus: carmine red [carmine].

Pupa: the intermediate stage between larva and adult; loosely applied for all orders, properly only for those with a complete metamorphosis: a pupa is obtect, when inclosed in a rigid case on which the members may or may not be outlined. It is liber, or free when the appendages are separately encased and there is no covering over the whole: see chrysalis.

Puparium -ia: in Diptera, the thickened larval skin within which the pupa is formed.

Pupate, Pupation: to become a pupa: the act of becoming a pupa.

Pupiferous: applied to that generation of plant lice which produces sexed individuals.

Pupigenous: =pupiparous; q.v.

Pupigerous: forming a larval pupariuni: coarctate: said of dipterous larva that contract to form an envelope for the inclosed pupa.

Pupil: the central mark of an ocellate spot.

Pupillate: spots or marks that have an eye-like centre.

Pupipara: a series of Diptera, in which the females do not extrude the young until they have reached the stage ready to pupate.

Pupiparous: bringing forth young ready to pupate.

Pupivorous: feeding upon pupa: especially applied to those Hymenoptera that are parasitic upon insects in the pupal stage.

Purpurascent: becoming purple in shade.

Purpureous -eus: purple [mauve].

Pustular a colored point of moderate circumference.

Pustulated hair: in Mallophaga those arising from unchitinized spaces.

Pygidium: the last dorsal segment of abdomen left exposed by the elytra: in Forficulidw, the last dorsal segment: in Diaspincr, the compound terminal segment.

Pygofer: the last segment of the abdomen in certain Homoptera, especially the lateral margins which appear in the ventral view; hence sometimes used in the plural - pygofers.

Pygophore: the large upper piece of the genitalia in Homoptera.

Pygotheca: the parts containing the genitalia in Homoptera.

Pyloric: referring to the posterior extremity of the chylific ventricle.

Pyloric valve: the specialized posterior portion of crop where there is no distinct gizzard.

Pyloric valvule: a circular projection of the stomach behind which is an enlargement of the intestine.

Pylorus: the chylific ventricle.

Pyriform: shaped like a pear.

Q

Quadra -ri -ro: as a prefix, means four.

Quadrate: square or nearly so.

Quadrifarium: having four rows.

Quadrilateral: four-sided: formed or bounded by four lines: in Odonata, a space on the wings of Zygoptera bounded by the lower sector of arculus, the submedian vein, a cross-vein between these two, and the lower part of arculus.

Queen: the actively reproducing female among worker insects.

Quiescent: not active: applied to the pupae in forms with complete metamorphosis.

Quiet: subdued: not conspicuous or contrasting in color or maculation.

R

Race: a variety of a species possessing constant characters which yet are not specific; usually occurring in a different faunal region from the type and may thus be geographical: nearly synonymous with subspecies.

Racemose: like a bunch of grapes: applied to ovaries when they form bunches or sacs.

Rachis: a ridge or keel dividing the spinning canal at base, in caterpillars; the shank of an antennal joint into which the lateral spines or other processes are inserted.

Radial: pertaining to the radius or radial vein.

Radial area: in Orthoptera; the space between the mediastinal (subcosta) and radial veins: see scapular area.

Radial cells: the wing area between the radius and media; often divided: in the plural (Comst.), are those cells anteriorly margined by the radius or its branches.

Radial cross vein: (Comst.), is that which divides cell, radius 1.

Radial sector: in general, the lower of the two primary divisions of the radius (Comst.): in Odonata, an indirect branch from the media, just below and parallel with media 2.

Radial vein: in Homoptera, the first important vein next the costa between it and ulnar: in Orthoptera, = radius (Comst.): in Diptera, = 2d longitudinal vein (Meigen), = radius 2 (Comst.).

Radiate veins: the longitudinal veins spreading fan-like in the anal field of secondaries: = anal veins; q.v.

Radiated: marked with lines proceeding from a common centre.

Radicle or Radicula: that joint of the antenna that is articulated to the head.

Radio-medial cross vein: connects the radial and medial systems and usually closes the radial cell (Comst.).

Radius: (Comst.); the third of the longitudinal veins starting from base and dividing into not more than five branches before reaching the margin: the branches are numbered 1, 2, 3, 4, and 5, respectively, beginning at the costal margin and extending outward and downward.

Radix: base of wings, and their point of insertion; see pteropega.

Rami -us: branches: a branch.

Ramification: the branching out in every direction.

Ramify: to branch out in every direction.

Rammel-kammer: = copulation chamber; q.v.

Ramose -ous: branched, or having long branches.

Rapacious: predatory; capturing and eating prey.

Raptatory: = raptorial; q.v.

Raptoria: applied to that series of Orthoptera, in which the anterior legs are fitted for grasping; Mantidae {*Scanner's comment: No longer Orthoptera, but Mantodea, a suborder of the Dictyoptera.*}

Raptorial: formed for seizing prey.

Rare: seldom seen or found.

Rasorial: formed for scratching; applied to leg structures.

Ravenous: greedy; voracious; hungrily.

Receptaculum seminis: a sac or pouch-like appendage at the junction of the oviducts with the vagina; it is filled during copulation and the eggs are fertilized from it as they are extruded.

Reclinate: directed backward; e.g. the bristles in Diptera.

Reclinatus: = reflexed; q.v.

Reclivate: curved into a convex, then into a concave line.

Recondite: the sting when concealed in the abdomen.

Rectal cauda: the terminal, tubular process or tail terminating the abdomen of some male Hemiptera.

Rectal glands: appendages to or thickenings of the rectum secreting a lubricating material.

Rectal tracheal gills: lamelliform structures in the rectum of the nymphs of some Odonata, supplied with trachea and tracheoles and serving as respiratory organs.

Rectangular: in the form of a right or rectangle.

Rectangulate: forming or meeting in a right angle.

Rectigrade: larvae which, having sixteen feet, walk with a rectilinear body.

Rectilinear: in the form of a straight line.

Rectum: a chamber, variable in size and form, just within the anus, in which the excretions are formed or molded for expulsion from the body:= cloaca.

Rectus: right or straight.

Recumbent: lying down; reclining.

Recurrent: running backward: applied to nerves it = stomatogastric.

Recurrent nervure: in Hymenoptera (Nort.), is the medial cross vein (Comst.), from the point of branching to the junction.

Recurrent vein: in Hemerobiidae, the first branch of the subcosta when it recurves toward the base of the wing.

Recurved: bowed backward.

Reductus: a zig-zag marking or corrugation.

Reflected or Reflexed: angularly bent backward.

Refracted: bent back as if broken.

Region: a space or area adjoining a specified point: a part of the body composed of a number of segments, as the head, the thorax, or the abdomen.

Rejuvenescence: a renewal of youth; bringing back to a condition of youth.

Remote: further removed than distant.

Reniform: kidney-shaped: applied to a macula approximating that shape, found at the end of median cell in many moths.

Repand: wavy; with alternate segments of circles and intervening angles.

Replicate: wings folded back upon the base; like the secondaries in Coleoptera.

Replicatile: capable of being folded back.

Repugnatorial: serving to repel: so offensive as to drive away: applied to glands that secrete an offensive material.

Reservoir: a case or cavity for the storage of any fluid or secretion.

Resilient: elastic; having the property of springing back.

Respiration: breathing or taking breath: union of oxygen with tissues and liberation of carbon dioxide from same.

Restricted: held back: confined to a limited area.

Resupinate: upside down; horizontally reversed.

Rete: the fatty mass of insects: also applied generally to any structureless membrane or layer.

Reticulate: like net-work.

Reticulum: a net-work; as of a cell.

Retina: that portion of the eye upon which the image is formed.

Retinaculum: in Lepidoptera, the loop into which the frenulum of the male is fitted; = hamus, q.v.: in Hymenoptera, horny, movable scales serving to move the sting or to prevent its being darted out too far: in Coleoptera, the middle, tooth-like process of the larval mandible.

Retinal pigment: the pigment layer of the compound eye just above the basilar or fenestrate membrane.

Retinophora: = retinula; q.v.

Retinula -ae: the retina of a single ocellus: the nerve fibres or cells between pigment cells and retina of the compound eye.

Retracted: drawn back; opposed to prominent.

Retractile: capable of being drawn in or retracted.

Retractor: used in drawing in or back; as a muscle.

Retroarcuate: curved backwards.

Retrocession: the going or moving backward.

Retrose: (sinuate), pointing backwards; (serrate) inversely serrated.

Retuse: ending in an obtuse sinus or broad, shallow notch, terminated by an obtuse hollow.

Reversed: turned in, an unusual or contrary direction, as upside down or inside out: said of wings when they are deflexed, the margin of secondaries projecting beyond those of primaries.

Reviviscence: coming back to life; awakening from hibernation.

Revolute: spirally rolled backward.

Rhabdites: the blade-like elements of the sting and ovipositor: a rod or bladelike process projecting from the epidermis.

Rhabdom: the rod lying in the axis of the retinula, below the crystalline cone of an eye.

Rhabdomere: the rod-like distal portion of a retinular cell.

Rhabdopoda: clasping organs of the 9th abdominal segment of male.

Rhinarium: a nostril piece or portion of the nasus: q.v.: in Odonata, the lower portion of clypeus = ante-clypeus; q.v.

Rhipiptera: = Strepsiptera q.v.

Rhomboidal: having the form of a rhomb.

Rhombus: a quadrangular figure having its four sides equal and its opposite lines parallel, with two opposite angles acute and two obtuse.

Rhopalocera: that series of Lepidoptera in which the antenna are alike in both sexes and form a club at tip.

Rhodoptera: apterous insects with sucking mouth structures.

Rhophoteira: an ordinal term for the fleas (Clairville).

Rhynchophora: that section of Coleoptera, in which the head is produced into a snout, at the end of which the mouth structures are situated; gular sutures confluent: prosternal sutures wanting: the weevils.

Rhynchota: = Rhyngota: q.v.

Rhynchus: of Fabricius, = promuscis: q.v.

Rhyngota: insects in which the mouth parts are prolonged into a beak or rostrum which serves as a protection to the piercing lancets: Hemiptera in the broad sense.

Rhythmical: occurring at regular intervals in the production of opposite conditions.

Rigid: inflexible: holding a direct course.

Rima: a crack or longitudinal opening with sharp edges.

Rimose: full of cracks.

Ring: a circle or annulus, usually margining a discolored spot.

Ringent: gaping.

Riparian: along the river or along shore.

Ripicolous: dwelling on river banks: riparian.

Rorulentum: dusty: =pulverulent: q.v.

Roseate-eous,-ens: rose colored: pale blood red [rose].

Rosette-shaped: in the form of a double rose: clustered round a centre.

Rostellum: a small beak; applied to the mouth parts of sucking lice: also used in Hemiptera, as = rostrum; q.v.

Rostral: pertaining or attached to a rostrum; specifically of Hemiptera.

Rostrate: the head, when it has a long protraction bearing the mouth parts.

Rostriform: produced like a beak or snout.

Rostrum: a snout-like prolongation of the head: in Coleoptera, applied to the rigid extension in the snout beetles: in Hemiptera, is the jointed beak covering the piercing lancets; and this is the better use of the term.

Rotate: wheel-shaped.

Rotator: used for turning; applied to muscles.

Rotatory: an articulation that permits a rotating motion, e.g. a ball and socket joint.

Rotula: a small round segment sometimes present between the joints of antenna; and palpi: = torquillus.

Rotule: the trochantine.

Rotundate: rounded: in the form of a circle or segment of a circle: without angles: said of margins when they pass gradually into each other and do not form an angle at their point of junction.

Royal jelly: the food supplied to bee larva that develop into queens.

Royal pairs or Royalties: the sexually active males and females of social insects.

Ruben: red, approaching carmine [alizar crimson].

Rubescent: reddish or becoming red.

Rubiginose: a rusty red [dragon's blood, but brighter].

Rubinous: ruby red: like the eye of a house fly.

Rudiment -any: the beginning of any structure or part before it has developed.

Rufescent: reddish.

Rufous: brick-red [chronic yellow + dragon's blood].

Ruga, Rugulae: a wrinkle; small wrinkles.

Rugose -ous: wrinkled: with irregular waved elevated lines.

Rugulose -ous; minutely wrinkled.

Runcinate: notched: cut into several transverse acute segments which point backward.

Ruptor ovi: = egg burster; q.v.

Rutilous: a shining bronze red.

S

Sabulose: sandy or gritty.

Sac: in Coccidae, the separate cottony envelope secreted by many species.

Saccate: gibbous or inflated toward one end.

Saccule: a little sac or pouch.

Saccus: a lobe of ventral plate of 9th segment in female Lepidoptera.

Saddle: the chitinous plate on the anal siphon of Culicid larvae.

Sagitta: arrow-like spots on the wings of Lepidoptera and other insects.

Sagittae: the inner pair of forceps in male genitalia of aculeate Hymenoptera: see stipites.

Sagittal: equivalent to longitudinal.

Sagittal plane: the longitudinal vertical plane which divides an animal into right and left halves.

Sagittate: shaped like an arrow head: elongate triangular.

Salient: projecting; jutting out.

Saliva: the secretion of the salivary glands that moistens and begins the digestion of the food.

Salivary glands: glands that open into the mouth or at the beginning of the alimentary canal, secreting a digestive, irritant or viscid material.

Salivary receptacle: a small cavity above the opening of the salivary duct, between labium and hypopharynx.

Salivary pump: applied to the chitinous, cup-like structure at the base of the labial stylets of piercing Diptera; e.g. mosquitoes.

Saltatoria: that series of Orthoptera in which the posterior legs are formed for jumping.

Saltatorial or Saltatory: formed for jumping or leaping: a posterior femur when much enlarged and dilated.

Saltatory appendage: in Collembola = furcula; q.v.

Sanguineous -eus: red like arterial blood [crimson lake].

Sanguinolent: bloody; in color or appearance.

Sapphyrinus: sapphire blue [French blue].

Sarcolemma: the elastic covering of the striated muscular fibres.

Sarothrum: the basal joint of posterior tarsus in pollen gatherers: see metatarsus.

Saturate: deeply or strongly marked; in a color, means intense.

Saxicolous: species that frequent rocky or stony areas.

Saws: the ovipositors of the saw flies or Tenthredinidae.

Scaber: uneven, rough.

Scabriculous: regularly and finely wrinkled.

Scabrose -ous: rough like a file, with small raised points.

Scalariform: ladder-like; applied to venation when the veinlets between two longitudinal veins are regularly arranged like the rungs of a ladder.

Scale: a general term to distinguish Coccidae: specifically the puparium of a Diaspid, comprising exuviae and excreted matter: the waxy covering of a male Lecaniid: in Diptera = alula: q.v.

Scales: broad flattened hairs, forming the wing vestiture of Lepidoptera, and present in various other insects.

Scalloped: an edge marked by segments of circles without intervening angles.

Scalpellus: a lancet-like piercing structure, as in some Diptera.

Scalpriform: chisel-shaped.

Scansorial: said of feet, when formed for climbing on hair.

Scape: the long basal joint of a geniculate antenna in Coleoptera; usually applied to the three basal joints, as in Hymenoptera.

Scaphiform: boat-shaped.

Scaphium: a ventral process of the 10th abdominal segment in male Lepidoptera below the uncus.

Scapula: in Lepidoptera the shoulder tippets, patagia or axillae; q.v.: in Hymenoptera, the side pieces of the mesonotum; also, a trochanter of the foreleg; in Proctotrupidae the lateral lobes on each side of the parapsidal furrow.

Scapulae: in Hemiptera, the inferior lateral face of mesonotum.

Scapular: the episternum; q.v.: applied to the scapula.

Scapular area: in a wing, is that portion nearest the shoulder: in Orthoptera, = radial area.

Scapularia: a meso-episternum: see scapula.

Scapular vein: in Orthoptera, = radius.

Scapus: = scape.

Scarabidoid: applied to that stage of a meloid larva in which it resembles a white grub or Scarabid larva.

Scarified: a surface with irregular depressions, as if clawed or scratched.

Scariose -ous: dry and scaly.

Scatophagous: feeding upon dung or excrement:= merdivorous.

Scent glands, or organs: glandular structures; sometimes eversible, sometimes in the form of hair tufts or pencils for diffusing odors that may be repellant or attractive; most frequently found in males as a secondary sexual character.

Scent pores: = ostioles; q.v.

Sclerite: any piece of the body wall bounded by sutures.

Scopa: a brush: a covering of short, stiff hair of equal length: in Hymenoptera, the thick hair covering the posterior tibia of pollen- gathering forms.

Scopate: furnished with a scopa.

Scopula: a small, dense tuft of hair: the bristles or stiff hairs covering the inner side of basal joint on the tarsi of pollen-gathering Hymenoptera.

Scopulipedes: bees which have pollen gathering structures on the feet.

Scraper: the hardened portion of the inner margin of the tegmina in crickets used in producing the song.

Scriptus: lettered or marked with characters resembling letters.

Scrobes: grooves formed for the reception or concealment of an appendage specifically, in Rhynchophora, grooves at the sides of the rostrum to receive the scape of antenna 2: also applied to grooves on the sides of mandibles: in Hymenoptera, the usually circular impressions upon the frons, in which the scapes revolve: in Orthoptera, the pits in which the antenna; are situate.

Scrobiculated: having the surface covered with deep round pits.

Scrotal membrane: the envelope covering the testes in some insects.

Scrotiform: purse-shaped.

Scrotum: = scrotal membrane; q, v.

Sculpture: the markings or pattern of impression or elevation on an elytra or other body surface.

Sculptured: a surface, when marked with elevations or depressions or both, arranged in some definite manner.

Scutate-iform: shield or buckler-shaped.

Scutcheon: = scutellum; q.v.: also used by some authors (Walker) for the pronotum in Homoptera.

Scutel: = scutellum: q.v.

Scutellar angle: of elytra is next to the scutel when wing is expanded.

Scutellar bridge: in Diptera, a small ridge on either side of the scutellum, connecting it with the mesonotum.

Scutellar space: in Mantids, an area between antennae and clypeus.

Scutellate: dish- or platter-shaped.

Scutellum: the third dorsal sclerite of the meso- and meta-thorax: in Coleoptera, the triangular piece between the elytra at base and universally referred to as the scutellum: in Heteroptera, a similar sclerite between base of hemelytra: in Diptera, a sub-hemispherical. body posteriorly cut off by an impressed line from the dorsum of the meso-thorax.

Scutes: the chitinous shields or plates on the segments of larvae.

Scutiform: = scutate; q.v.

Scutum: the second dorsal sclerite of the meso- and meta-thorax.

Seal-brown: a brilliant deep red-brown: almost like, but darker than castaneous [between dragon's blood and Indian red].

Sebaceous: fatty or oily; applied to glands secreting such substances.

Sebific: oily; sebaceous; somewhat sticky.

Sebific duct: carries the excretions of the colleterial gland to the bursa copulatrix.

Second antennal segment: the third or tritocerebral segment of head.

Secondaries: the hind wings; always attached to the meta-thorax.

Secondary sexual characters: features possessed by one sex but not the other, other than the differences of the reproductive organs themselves; e.g. color, size, shape, etc.

Second clypeus: see anteclypeus.

Second costal cell: in Hymenoptera (Pack.); is the stigma.

Second inner apical: in Hymenoptera (Nort.), = media 3 (Comst), to the junction of medial cross-vein; also called sub-marginal nervure in part.

Second lateral thoracic suture: in Odonata, extends front base of secondaries to the rear of the third coxa.

Second longitudinal vein: in Diptera:= radius 2 + 3 (Comst). Second maxilla: the labium: q.v.

Second maxillary segment: the seventh or labial segment of head.

Second median area: see median area; areola.

Second submarginal nervure: Hymenoptera (Nort.), = radius 5 (Comst.).

Secretion: any matter produced by a gland: in Coccidae, specifically, the waxy, fibrous, cottony or silky substances forming the "scales."

Secretionary supplement: that part of a Diaspid scale extending beyond or around the pellicles.

Secretory: concerned in the process of secretion.

Sectores coronis: the tearing or cutting structures used by the Lepidoptera in working out of a cocoon.

Sectors: longitudinal veins in Odonata, which strike the principal veins at an angle, and usually reach the apex or hind margin: they are radial, subnodal, principal, nodal, median, short, and upper and lower of triangle: all of which see.

Secund: pointing one way: unilateral.

Securiform: triangular-compressed; like the blade of a hatchet.

Sedentary: not active: settled or remaining in one place.

Segment: a ring or division bounded by incisions or sutures: a segment of an insect or of any articulate is a transverse portion reaching entirely across the body, originally separated on the exterior by incisions or sutures from the preceding and the succeeding segments, having attached to it not more than one pair of ventral appendages, containing internally not more than one pair of nerve ganglia which supply nerves to the pair of appendages; = somite, arthromere: fusion of segments frequently obscures, as in the head: externally the walls of one segment may be composed of a number of sclerites separated from each other by secondary sutures.

Segmentate: made up of rings or segments.

Segmentation of egg: the division of the originally single celled egg into a number of coherent cells or blastomeres; = cleavage.

Segregated: detached or scattered into groups.

Segregation: a separation or placing apart.

Sejunctus: separated.

Sellate: saddle-shaped.

Sematophore: a seminal packet, composed of the seminal fluid mixed with the excretions of the accessory glands.

Sembling: = assembling; q.v.

Semen: the fluid secreted in the testes, containing the spermatozoa.

Semi-: half.

Semicircular: like the half of a circle.

Semi-complete: in metamorphosis, = incomplete; q.v.

Semicordate: half or partly heart-shaped.

Semicoronate: partly surrounded by a margin of spines, hooks or the like.

Semicoronet: a margin of spines or hooks partly surrounding a structure or process.

Semi-cylindrical: like a groove or half a cylinder.

Semi-hyaline: hyaline in part only: not altogether transparent.

Semi-looper: a caterpillar in which one or two pairs only of the abdominal legs are wanting and where in progression, only small loops are formed: see looper.

Semi-lunar: in the form of half a crescent.

Semi-lunar valve: guards the auriculo-ventricular opening of the heart.

Seminal ducts: = vasa deferentia; q.v.

Seminal vesicles: enlarged tube or pouch-like structures which serve to store the seminal fluid of the male, and in which the later stages of its development may take place.

Seminiferous: semen-secreting.

Semipupa: that stage of the larva just preceding pupation: more specifically the interpolated stage between the active larva and the true pupa, in hypermetamorphosis.

Semi-saggitate: like the longitudinal half of an arrow head.

Semitropical or Gulf strip: is the southern part of the Austro-riparian area extends from Texas to Southern Florida, covers a narrow strip in So. Georgia and probably follows the coastal lowlands into South Carolina.

Sensim: gradually.

Sensoria: the circular openings covered by membrane, on the antenna or legs of plant lice.

Sensory: relating to or having a sense function.

Sensory pittings: deep pits or punctures through the surface, which may or may not bear pegs, bristles or seta, and may be open or covered by a membrane; serving as organs of perception for sounds or smells.

Septa, Septula: in Odonata, the triangular area of the mesonotum before the insertion of the primaries: = calli axillary.

Septum: an internal division of a body cavity.

Sequence: the order in which things follow; e.g. species or genera.

Seriatim: placed in longitudinal rows.

Sericeous: silky: clothed with very dense minute hair which gives a silky lustre.

Sericterium -ies: the silk producing gland or glands in caterpillars: the spinning structures.

Series: a group of species, genera or families, arranged to show agreement in a common character which is not of sufficient importance to warrant the next higher division.

Serific glands: are these which produce a thick, mucous-like secretion which, on hardening, forms silk.

Serosa: the outer membrane that envelops the forming embryo, the amnion and the remainder of the egg.

Serpentinous: a dirty, dark green [Hooker's green].

Serra: a saw or saw-like part.

Serrate: saw-toothed, the teeth set toward one end.

Serrato-dentate: toothed, the edges themselves saw-toothed.

Serratulate: with little teeth or serrations.

Serricornia: that series of Coleoptera in which the antenna are serrate or saw-toothed.

Serriferous: possessing a saw-like ovipositor in the female; the saw- flies.

Serrulate: with numerous little saw teeth.

Serum: the fluid in which the blood corpuscles float or are suspended.

Sesquialter or Sesquiocellus: a large ocellus including a smaller one.

Sesquitertial: occupying a fourth part.

Sessile: closely seated: the abdomen, when it is closely attached for nearly or quite its full width to the thorax.

Sessiliventres: Hymenoptera in which the abdomen is sessile.

Seta -ae: a pointed bristle or long stiff hair: slender, hair-like appendages.

Setaceous: bristle-shaped: slender, gradually tapering to a tip.

Setarious aristate: the dipterous antenna when the arista is simple.

Setiferous: = setigercus; q.v.

Setiform: in the form of a bristle or seta: when a slender short bristle arises from a thicker basal joint.

Setigenous: the hypodermal cells that give rise to setae.

Setigerous: bearing setae or bristles; e.g. punctures.

Setiparous: producing hair or sets.

Setireme: the hairy, oar-like legs of aquatic insects.

Setose -ous: bristly or set with bristles.

Setula: a small stiff bristle or seta: in Diptera, the small thorn at the end of the subcosta.

Setulose: clothed with fine seta or setulae.

Sex: as a number, six: the physical difference between male and female: usually indicated by the sign of Mars for male, and Venus for female; workers or undeveloped females have the sign of Venus without the cross line, or a combination of the two others.

Sexuparae: that generation of plant lice which produces the true sexes.

Shade: a cloudy, ill-defined streak or band.

Shagreened: a surface roughened with minute tooth-like projections.

Shank: = tibia; q.v.

Shard: a chitinous sheath or elytron.

Sharp: with a pointed tip or thin edge; opposed to blunt.

Sheath of penis: in Odonata, a median, hood-like piece between the hamules, under which the penis is folded when not in use.

Shin: = tibia; q.v.

Short sector: in Odonata, = media 4 (Comst.).

Shoulder: loosely applied to an obtuse angulation; more generally to the humeral angle of fore wings or elytra: the anterior angles of thorax in Lepidoptera; the angles of prothorax in Heteroptera: the lateral angles of metazona of pronotum in Orthoptera.

Sialisterium: a salivary gland.

Side: the lateral margin of the body.

Side piece: in genitalia of male Culicids the main lateral part of the clasping organ or basal segment of clasp.

Sides of thorax: in Odonata, includes the pleura of meso- and meta- thorax, less the meso-episterna.

Sienna: a brownish orange [brown ochre].

Sigmoid: shaped like the Greek letter sigma, or English S.

Signate -us: = with marks or spots; see notate.

Signature: a colored blotch of any size or shape.

Silaceous: = ochraceous.

Silk: the hardened salivary secretion of certain larvae, mainly of Lepidoptera. similar material is produced by anal glands of some larva in Neuroptera.

Silk-glands: a pair of modified salivary glands in certain larva, mostly of Lepidoptera that secrete a viscid fluid which, on contact with the air, hardens into a silken fibre.

Silvicolous: living in moist, shady woods.

Simple, Simplex: without process, armature, or appendage of any kind.

Simple eyes: = ocelli; q.v.

Sinciput: in Coleoptera; that part of the vertex between the eyes.

Sinistrad: toward the left.

Sinistral: extending to or at the left from the median line.

Sinistro-caudad: extending obliquely from the left toward the tail.

Sinistro-cephalad: extending obliquely from the left toward the head.

Sinuate: cut into sinuses; applied to lines and margins with an in and out curve.

Sinuated: winding: with the edge scooped into sinuses.

Sinuato-convex: sinuate and convex.

Sinuato-lobate: sinuate and lobed.

Sinuato-truncate: truncated, with the margin sinuate.

Sinuous: undulating; curved in and out.

Sinus: a curvilinear indentation more or less profound: an excavation as if scooped out: a curved break in an otherwise straight margin.

Siphon: a tube-like mouth organ in certain insects: the breathing tube of a Culicid larva: any tubular external process or structure.

Siphonaptera: an ordinal name for insects which are wingless: mouth formed for piercing and sucking; saltatorial; transformations complete: the fleas = Aphaniptera; q.v.

Siphonata: = Homoptera or, more specifically, plant lice and leaf hoppers.

Siphonets: see honey tubes.

Siphonophora: = Coccinellidae; the term is preoccupied in the Coelenterates.

Siphunculata: the sucking lice.

Siphunculus: the suctorial organ of a louse, contained within the tubule: in plant lice =honey tubes; q.v.

Situ (in): in its natural place or position.

Sixth longitudinal vein: in Diptera; = 1st anal vein (Comst.).

Skeleton: the hard chitinous parts which externally (exoskeleton) or internally (endoskeleton) form a protective covering, or serve as points of attachment, to muscles and other soft organs.

Skippers: a popular term for Hesperid butterflies: the dipterous larva sometimes found in cheese and other provisions.

Slaty: very dark blackish gray with a reddish tinge [neutral with a little Indian red].

Slug: in general, any larva that has a slimy viscid appearance, and the body closely applied to the food plant: more specifically, the larvae of certain sawflies and of some Coleoptera.

S. M. interspace: sub-median interspace in the primaries of Lepidoptera, includes the space between the median and sub-median veins; (cubitus and 1st anal, Comst.).

Smaltinus: a dull grayish blue.

Smaragdinus: emerald green [pale green].

Smooth: a surface without elevations or indentations.

Snout: the prolongation of the head in Rhynchophora at the end of which the mouth parts are situated: see rostrum.

Social: living in communities: more especially those species in which undeveloped or worker forms occur and where the colony has a single female head.

Soldiers: in termites; forms sexually undeveloped, in which the mandibles are pincer-like and the head is much enlarged: worker majors in certain ants.

Solid: applied to an organ usually jointed, when these joints form into one mass; e.g. the capitulum of certain clavate antennae.

Solitary-arius: occurring singly or in pairs; not in colonies.

Somatic: relating to the body, or abdomen.

Somatotheca: that part of pupa covering abdominal rings:= gasterotheca.

Somite: = arthromere.

Sonifaction: the production of sound: = stridulation; q.v.

Sonoran faunal areas: see upper and lower Sonoran.

Sonorific: sound producing: applied to stridulating organs.

Sordid: dirty; dull.

Spadiceous: bay brown [dragon's blood + brown ochre].

Spado: the worker or neuter in bees and ants.

Sparse: scattered: single hairs, scales or sculptures set well apart.

Spatha: a median piece in male genitalia of aculeate Hymenoptera, covering the bases of the sagitte.

Spatula: the breast bone (q.v.) of cecidomyid larvae.

Spatulate: rounded and broad at top, attenuate at base.

Specialization: the adaptation of an organ to a definite purpose, or of an organism to fit a determinate environment.

Species: an aggregation of individuals alike in appearance and structure, mating freely and producing young that themselves mate freely and bear fertile offspring resembling each other and their parents: a species includes all its varieties and races.

Specific character: a feature common to all individuals of a species, by means of which they may be distinguished from all other individuals of other species: = essential character.

Specular: mirror-like: transparent.

Specular membrane: in male Cicada, the inner or posterior mirror-like membrane of the sound-organ: = mirror.

Speculum: a transparent area or spot on wings of some Lepidoptera; the glassy areas at base of tegmina in male Orthoptera that serve as sounding boards: a spot on the neck of some caterpillars.

Sperm: the seminal fluid: in plural form is sometimes used as = spermatozoa; q.v.

Spermatheca: = spermatotheca; q.v.

Spermatid: the final cells which are converted without further division into spermatozoa: they arise by division of the second spermatocytes (Wilson).

Spermatocytes: the cells arising from the spermatogonia. The primary spermatocyte arises by growth of one of the last generation of spermatogonia. By its division are formed two secondary spermatocytes, each of which give rise to two spermatids (Wilson).

Spermatogenesis: the development of spermatozoa.

Spermatogonia: the descendants of the primordial germ cells in the male. Each ultimate spermatogonium typically gives rise to four spermatozoa.

Spermatophora: a sac or case containing spermatozoa.

Spermatotheca: the sac or reservoir in the female, that receives the sperm during coition: = spermatheca and receptaculum seminis.

Spermatozoön-zoa: the male cell or cells which, by uniting with the ova, fertilize them.

Spherical: in the form of a sphere: a body in which all diameters are equal.

Spherule: a minute sphere or globule.

Spicula: a slender needle-like process: e.g. the sting in bees: also employed as = ovipositor: q.v.

Spiculiform: like a slender, needle-like process.

Spiculum: a small spicule or thin, pointed process.

Spinate: produced into an acuminate spine.

Spindle-shaped: cylindrical, elongate, thicker in the middle, tapering to each end: fusiform.

Spine: a sharp process: in Coccidae there are two, one each side of each segment of the pygidium.

Spiniferous: bearing, or clothed with spines.

Spiniform: in the form or shape of a spine.

Spinneret: the ligula in bombycid and some other larvae, modified for silk spinning: any organ consisting of an internal tube, terminating in a pore, spine or process, producing a silky or waxy fibre: in the plural, the organs concerned in the emission of the silky or cottony filaments of which the scales or sacs of Coccidae are produced: = fusulus.

Spinose -ous -ed: set with acute processes or spines.

Spinous-radiate: beset with spines in a circle, either concatenate, united at their bases, or setaceous, like bristles.

Spinulae: spinous processes at the apex of the tibia: also called spines, spurs or heels.

Spinulate -ose: set with little spines or spinules.

Spinules: little spines.

Spira: the coiled ovipositor of Cynipidae.

Spiracle -cula: a breathing pore: q.v.: in the plural the lateral openings on the segments of the insect body through which air enters the tracheae:= stigmata.

Spiracular area: the anterior of the three areas between lateral and pleural carinae on the metanotum of some Hymenoptera: =first pleural area.

Spiracular line: in caterpillars is that which includes the spiracles: = stigmatal.

Spiracular sulcus: on the metanotum of Hymenoptera, is a grooved linear channel extending from spiracle to apical margin.

Spiral: rolled up like a watch spring, or twisted like a cork-screw.

Spiral fibre: the spiral thickening or folding of the chitinous lining of a trachea, which gives to the latter its characteristic microscopic appearance as well as its support and elasticity:= ctenidium.

Spirignath: = spiritrompe.

Spiritrompe: the spiral tongue in Lepidoptera.

Splanchnic: applied to the outer embryonic layer of the rudiment of the mid-intestine: or the inner layer of the mesoderm which, becoming applied to the walls of the alimentary canal, develops into the muscle fibres thereof.

Splendens -ent: shining: with a metallic glitter.

Splitter: one who splits or describes species or genera upon minute characters which the "Lumper" (q.v.) deems insufficient to authorize them.

Spongioplasm: the net-like structure of protoplasm in a cell.

Spongiose: a soft, elastic tissue resembling a sponge.

Spoon: = bouton; q.v.

Spring: in Collembola. = furcula: q.v.

Spur: a short, stiff, generally blunt process and usually not articulated at its base: in the plural refers to paired spiniform processes at or near the end of tibia:.

Spur formula: a numerical expression of their arrangement; as 2-3-4; two spurs on fore tibia, 3 on middle, and 4 on posterior; in Trichoptera.

Spurius: false: applied to aborted anterior legs in some diurnal Lepidoptera. Spurious cell: in Diptera (Pack.), = 3d anal (Comst.).

Spurious veins: certain folds or thickenings in the wing surface which resemble a vein so nearly as to be readily mistaken and sufficiently constant to be useful in classification.

Squama: in Odonata, the sclerite that bears the palpus of both maxilla and labium: the scale-like first abdominal segment of some ants: a scale-like appendage covering the base of primaries in Lepidoptera, and so = patagium; q.v.: a small scale above the halteres in Diptera: in this order Packard uses squama for the lobed scale and restricts alula to the lobe-like appendage: Osten-Sacken uses squama for the posterior scale alone and antisquama for the anterior.

Squames: the flattened, fimbriated or spine-like marginal processes of the pygidium in Diaspinae, other than the lobes and true spines: = "plates," of Comst.; or "scaly hairs" of Maskell.

Squamiform: having a scale-like form.

Squamose -ous: scaly or covered with scales.

Squamula: a small corneous scale covering the base of primaries in some insects: = tegula, q.v.: in Diptera, = alula.

Squarrous: scurfy: clothed with rough scales differing in direction, standing upright, or not parallel to the surface.

Stadium -ia: the interval between the molts of larvae:= instar q.v.: any one period in the development of an insect.

Stage: refers to the period of development; e.g. larval, pupal, etc.

Stalked bodies: = gyri cerebrales; q.v.

Stellate: star-shaped; with four or five radiating lines.

Stelocyttares: social wasps in which the comb layers of the nest are supported by pillars and not connected with the envelope: see poecilocyttares and phragmocyttares.

Stemapoda: the modified filamentous anal legs of Cerura and other Notodontid larvae.

Stemmata: simple eyes or ocelli; q.v.

Stem-mother: in plant lice; that form hatching from the winter egg, which starts a series of agamic summer generations.

Stenocephalous: with a narrow, elongate head.

Stenorhynchan: narrow beaked or snouted.

Stenothorax: a supposed ring between pro- and meso-thorax.

Stercoral: relating or pertaining to excrement.

Sterile: not capable of reproducing its kind.

Sternal orifice: in Perlids; a peculiar slit on each side of the sternum, extending inward and ending blindly := furcal orifice.

Sternal spatula: = breast-bone; q.v.

Sternauli: the short and often obsolete furrows on either side of the mesosternum in Hymenoptera.

Sternellum: the second sclerite of the ventral part of each thoracic segment frequently divided into longitudinal parts which may be widely separated.

Sternite: the ventral piece in a ring or segment.

Sternopleura: in Diptera, the lower part of the pleura, below the sternopleural suture and above the front coxa.

Sternopleural bristles: in Diptera, are situated on the sternopleura below the sternopleural suture.

Sternopleural suture: in Diptera, is below and nearly parallel with dorso-pleural suture, separating the mesopleura from the sternopleura.

Sterno-rhabdite: a sternal rhabdite.

Sternorhynchi: that series of Homoptera in which the beak or rostrum apparently arises from the sternum between the anterior coxae; e.g. plant lice, etc.

Sternum -a: the breast: the middle portion of the under surface of thorax, between the coxal cavities.

Stethidium: the trunk: the entire thorax with all its appendages.

Stigma: a spiracle or breathing pore: a dense, often discolored portion of the costal margin of a wing, usually at the end of the radius; see anastomosis: in Diptera, a colored wing spot near the tip of the auxiliary vein: in Lepidoptera, the specialized patch of black scales on the primaries of Hesperidae.

Stigmata: the spiracles: also applied to the two spots, orbicular and reniform, in the cell of the primaries of certain moths.

Stigmatal line: in caterpillars, = spiracular; q.v.

Stigmatiferous: applied to processes or structures bearing spiracles or stigmata.

Stimuli: the small acute spines on some larva, especially wood-borers

Sting: the modified ovipositor in aculeate Hymenoptera.

Stipes: the foot-stalk of the maxilla; articulated partly to the head, partly to the cardo, and bearing the movable parts: modified into a piercing structure in some Diptera and into a lever for flexing the proboscis in others.

Stipitate: supported on a stalk or pedicle.

Stipites: the outer pair of forceps in male genitalia of aculeate Hymenoptera see sagittae.

S. T. line: sub-terminal line; crosses the primaries of many moths just before the outer margin.

S. T. space: the area between the t.p. line and s. t. line in moths.

Stirps: a stock or stem: a division of classification similar to super-family: not used at present.

Stoma -ata: a breathing pore or pores:= stigma; q.v.

Stomach: that portion of the alimentary canal, immediately following the gizzard and preceding the ileum, into which most of the digestive juices are poured = chylific ventricle.

Stomatodaeum: that invagination of the ectoderm that forms the mouth, pharynx, oesophagus, crop and gizzard.

Stomatogastric: that system of nerves and ganglia, lying along the dorsal and lateral surfaces of gullet and crop.

Stomatotheca: that part of the pupa covering the mouth structures.

Stramineous -eus: straw yellow [pale clay yellow].

Strangulate: constricted, as if by bands or cords.

Stratified: arranged or made up in layers.

Strepsiptera: twisted-wing: an ordinal term proposed for the parasitic Stylopidae, now ranged as a family of Coleoptera = Rhipiptera.

Stria: in Coleoptera, a longitudinal depressed line or furrow, frequently punctured, extending from base to apex of elytra: in Lepidoptera, a fine transverse line: in general, any longitudinal impressed line.

Striate -ed: marked with parallel, fine, impressed lines; or, in Lepidoptera, with numerous fine transverse lines.

Stridulate: to make a creaking noise by rubbing together two ridged or roughened surfaces.

Stridulation: a creaking sound produced by rubbing together two striated or otherwise roughened surfaces: the act of stridulating or the noise produced by it.

Striga: a narrow, transverse line or slender streak, either surface or impressed. {*Scanner's note: the proper plural is strigae*}

Strigate: having striga: applied to a surface on which the striga are impressed as in the elytra of some beetles, or to an ornamentation composed of fine, short lines.

Strigile -is: maculation that consists of parallel longitudinal lines: a deep sinus near base of first joint of anterior tarsus.

Strigillate -ation: = stridulate -anon; q.v.

Strigose: clothed with rigid bristles that are thickest at base: rough, with sharp bristles: = hispid.

Strigula: a fine short transverse mark or line.

Strigulated: with numerous strigulae.

Striolate -us: with finely impressed parallel lines.

Stripe: a longitudinal streak of color different from the ground.

Style: in Aphids, the slender tubular process at the end of the abdomen: in Coccids, a long spine-like appendage at the end of the abdomen of the male; = genital spike: in Diptera, the ovipositor (Loew); the single immovable organ immediately below the forceps in male Tipulidae (O-S.) a thickened jointed arista at or near the tip of the third antennal joint in the plural form applied to small, usually pointed, exarticulate appendages, most frequently found on the terminal segments of abdomen.

Stylet: a small style or stiff process: one of the piercing mouth structures in Diptera and Hemiptera.

Styliform: in the shape of a stylus: terminating in a long slender point, like the antenna in some Diptera.

Stylopized: infested by a member of the Stylopidae.

Stylotrachealis: with a long tube bearing a stigma, from the head case; as the pupa of some Diptera.

Stylus: a small, pointed, non-articulated process.

Sub-: as a prefix, means that the main term is not entirely applicable, but must be understood as modified in some way; e.g. sub-ovate, may be either more or less than ovate and may be irregular in outline.

Sub-aduncate: somewhat hooked or curved.

Sub-anal plate: Orthoptera; = sub-genital lamina; q.v.

Sub-apical lobe: of male genitalia in Culicids is the inner sub-apical lobe of the side piece.

Sub-apterous: almost wingless; with rudimentary wings only.

Sub-clavate: somewhat thickened toward tip; but not quite club- shaped.

Sub-coriaceous: somewhat leathery.

Sub-cortical: beneath the bark; as in larval borings, etc.

Subcosta: (Comst.); that longitudinal vein extending parallel to the costa and reaching the outer margin before the apex; not branched as a rule of Packard, in Hymenoptera, = radius (Comst.).

Subcostal cell: in Diptera (Schiner), = marginal cell (Loew), = radial 1 (Comst.) in the plural (Comst.), all those cells anteriorly margined by the subcosta first s.c. cell in Hymenoptera (Pack.), = radial and first radial 1 (Comst.).

Subcostal crossveins: in Odonata, are between subcosta and media on the basal side of the first antecubital.

Subcostal fold or furrow: lies between costa and radius.

Subcostal nervule: Lepidoptera, on secondaries:= media 1 (Comst.): s.c. 1 = radius 1 (Comst.); s.c. 2:= radius 2 (Comst.) s.c. 3 = radius 3 (Comst) s.c. 4 = radius 4 (Comst.) s.c. 5 = radius 5 (Comst.).

Subcostal vein: in Diptera (Schiner), = 1st longitudinal vein (Meigen) radius 1 (Comst.): in Lepidoptera, runs from base, parallel to costa, to or beyond the middle, giving rise to branches which extend to the outer margin and thus = radius (Comst.).

Sub-cristate: with a moderately elevated ridge or keel on pronotum, in Orthoptera.

Subcutaneous: under the skin: applied to larvae that feed under the skin of animals or within the substance of a leaf.

Sub-dorsal: the space between the dorsum and the stigmata.

Sub-dorsal line: in caterpillars is to the side of the dorsal and between it and the lateral or, if there is an addorsal line, between that and the lateral.

Sub-dorsal ridge: in slug caterpillars, extends longitudinally along the sub-dorsal row of abdominal tubercles.

Sub-equal: similar, but not quite equal in size, form or other characters.

Sub-eroded: wing margins when somewhat, but irregularly, indented.

Sub-falcate: when a wing is only a little excavated below the apex.

Subfamily: a division of classification containing a group of closely allied genera; different from other allied groups, yet not so as to make a family series: opinionative, and ending in -inae.

Sub-fossorial: legs used in digging: yet not greatly modified.

Sub-frontal: close to the front; immediately behind the front margin.

Sub-fulcrum: a sclerite between mentum and palpiger: rarely present.

Sub-fusiform: somewhat spindle-shaped.

Subgalea: a maxillary sclerite or segment, attached to the stipes, and bearing the galea or outer lobe.

Sub-geniculate: applied to antennae that are articulated from a short, thick scope.

Subgenital lamina or plates: plates underlying the genital organs in Orthoptera.

Subgenus: a division within a genus, based upon a character not sufficient for generic separation; opinionative.

Subglossa: in Odonata; a sclerite between the two halves of the mentum (Graber): is really the true mentum.

Sub-imago: sometimes applied as = nymph: that stage in Ephemerida just after emergence from the pupa and before the final molt during flight: that stage in the development of insects with free pupa when the insect is fully colored but yet retains its pupal position.

Sub-labrum: =epipharynx; q.v.

Sublingual: beneath the tongue; applied to a pair of salivary glands in bees.

Submargin -al: an imaginary portion of a surface outside of the disk and within the margin: a line is submarginal when it is well within the margin but close to it.

Submarginal area: of secondaries, lies between the costal margin and the 1st strong vein.

Submarginal cells: in Hymenoptera (Norton) = radial cells (Comst.): in Diptera (Will.); = radial 3 (Comst.).

Submarginal nervure: in Hymenoptera (Nort.); the irregular line of veins extending on the whole parallel with the outer margin; composed in part of media 1, 2, 3 and 4, the medial cross vein and cubitus 1 (Comst.).

Submedian cells: in Hymenoptera (Pack.); 1st = cubital + cubital 1 (Comst.) 2d = medial 3 (Comst.); 3d = 2d medial 2 (Comst.).

Submedian vein: in Odonata, =cubitus (Comst.); in Lepidoptera, = 1st anal (Comst.), runs from base of primaries to the hind angle, close to the inner margin and is v. 1 of the numerical series.

Submental: pertaining to the submentum.

Submental peduncle: in Coleoptera, the prolonged portion of the gula supporting the mentum.

Submentum: the basal sclerite of the labium, by means of which it is attached to the head.

Subnodal sector: in Odonata, = radial sector (Comst.).

Sub-nymph: applied to the resting or pupal stage of female Coccidae; also to a supernumerary stage before the formation of the pupa, and thus = pseudopupa.

Sub-ocellate: an ocellate spot that is blind or without a pupil.

Sub-ocular: beneath or below the eyes.

Sub-oesophageal ganglion: situated in the head below the oesophagus, formed by a union of the posterior three primitive head ganglia.

Suborder: a division of an order higher than a family, based on a character common to a large series of species; e.g. the Homoptera and Heteroptera in the order of Hemiptera.

Sub-pedunculate: in Coleoptera, when the constriction between pro- and mesothorax is so great as to give the appearance of a narrow waist.

Subreniform: a rounded spot or outline, below and sometimes attached to the reniform spot in Catocala and some allied Noctuids.

Sub-parallel: nearly parallel.

Sub-primary sub-ventral tubercle: on the thoracic and abdominal segments of caterpillars; sub-ventral, posterior, not present in the primitive first stage; it is VI of the abdomen, V of the thorax: constant.

Sub-sellate: nearly like or approaching the form of a saddle.

Sub-servate: denticulate.

Subspecies: a well-marked form of a species differing from the type in some character of color or maculation which is recognizable but does not prevent a fertile union: an indefinite and opinionative division.

Subspiracular line: in caterpillars, margins the spiracles inferiorly.

Substigmatal: that portion of the marginal cell below the stigma, in bees: = 1st radial 1 (Comst.): applied to a line in caterpillars = subspiracular.

Sub-teres: nearly but not quite cylindrical.

Subtile -is: slightly; feebly; small; pretty; graceful.

Subtriangular space: = internal triangle: see triangle.

Subtus: beneath; at the under surface.

Subulate: awl-shaped: linear at base, attenuate at tip.

Subulicornia: with awl-shaped antennae; applied to a combination of Odonata and Ephemerida.

Subuliform: formed like an awl: = subulate.

Sub-ventral line: in caterpillars, extends along the sides just above the base of the feet at lie edge between lateral and ventral.

Sub-ventral ridge: in slug caterpillars extends longitudinally along the sub-ventral series of abdominal tubercles.

Sub-ventral space: in slug caterpillars is the area on each side, between the lateral ridge and the lower edge of the body, and contains the spiracles.

Succincti: those chrysalids of butterflies which are held in place by a silken cord passing around the body: see suspensi.

Succineous: resembling amber in color or appearance.

Sucking pump: in sucking insects, a thick-walled muscular enlargement of the oesophagus that serves to draw up the liquid food = pharyngeal pump.

Sucking spears: the mandibles and maxillae of Hemerobiid larvae, used for puncturing prey and sucking its juices.

Sucking stomach: a thin-walled muscular pouch connected with the end of the oesophagus; serves as a food reservoir and is not commonly present except in some Lepidoptera.

Suctoria: an ordinal term proposed for fleas.

Suctorial: adapted for sucking: see haustellate.

Suctorial vesicles: bladder-like structures connected with the oesophagus in mosquitoes supposed to assist in blood-sucking; but this is disputed.

Suffused: clouded or obscured by a darker color.

Suffusion: a clouding, or a spreading of one shade over another.

Sulca: grooves, furrows or channels: plural of sulcus.

Sulcated: grooved; furrowed with broad, concave, parallel impressed lines.

Sulciform: resembling a sulcus.

Sulcus: a furrow or groove: a groove-like excavation.

Sulphureous -eus: bright, sulphur yellow [chrome lemon].

Superans: exceeding in size and length.

Superciliary: placed above the eyes.

Supercilium: an arched line over an ocellate spot.

Super-family: a division of classification less than an order, including a series of family groups more closely related to each other than to similar groups within the order: opinionative and ending in oidea: sometimes hardly different from suborder; but lower than suborder when both terms are employed.

Superficies: the upper surface.

Supericornia: those Heteroptera having the antenna inserted on the upper parts of the sides of the head; e.g. Coreidae: see infericornia.

Superior wings: the primaries; q.v.

Superlinguae: the lateral pair of organs of hypo-pharynx in Thysanura.

Superlingual segment: the fifth segment of head.

Superne: denotes all those parts belonging to the upper surface.

Supernumerary: additional or added cells, veins or other structures.

Supernumerary segment: in Cecidomyidae, between the head and first thoracic segment.

Super-order: a group of allied orders, like the Linnaean Neuroptera.

Superposed: placed one above the other, as the frontal tufts in some moths.

Supplementary sectors: interposed sectors; q.v.

Suppression: the non-development of a part normally present.

Supra-: over; above.

Supra-alar bristles: in Diptera, are situated, one on the post-alar callus, one on the alar frenum, the third on the edge of the supra-alar depression.

Supra-alar cavity: = supra-alar groove.

Supra-alar depression: in Diptera = supra-alar groove.

Supra-alar groove: in Hymenoptera, a groove or depression just above the base of wings: in Diptera, a groove on the meso-thorax just above the root of the wings.

Supra-anal: situated above the anus.

Supra-anal hook: in male of most Lepidoptera, a curved hook attached to the plate covering the genital cavity: = uncus.

Supra-anal plate: a triangular sclerite covering the anal cavity above; present in many insects, sometimes in one sex only, often in both: see anal operculum.

Supra-cerebral: applied to that pair of salivary glands situated above the brain in bees.

Supra-clypeal mark: in bees; a patch of light color above the clypeus.

Supra-clypeus: = post-clypeus; q.v.: = nasus.

Supra-Oesophageal: situated above the oesophagus: applied to two large ovoid ganglia so situated, and connected by a short, thick commissure; - the brain.

Supra-orbital: situated above the eye.

Supra-spinal: above the spine or nerve cord: applied to a cord or band of connective tissue lying above the central nervous system in adult Lepidoptera also to a sinus or vessel acting as a ventral heart.

Supra-spiracular line: in caterpillars, margins the spiracles superiorly.

Supra-stigmatal line: = supra-spiracular lines.

Supra-triangular cross-veins: in Odonata, cross the supra-triangular space.

Supra-triangular space: in Anisoptera, an area just above the triangle, occupying nearly the same position as the quadrilateral of Zygoptera: hyper-trigonal space.

Suranal: = supra-anal.

Suranal plate: the middle dorsal plate attached to the l0th abdominal segment of the male grasshoppers, above the anal opening: a supra- anal tergite of a caterpillar.

Sursum: directed upwardly.

Suspensi: the chrysalids of butterflies that are suspended by the tail only: see succincti.

Suspensoria: are those muscles or ligaments that hold the viscera and other internal structures in place.

Sustentors: the two posterior projections of a butterfly chrysalis.

Suture: a seam or impressed line indicating the division of distinct parts of body wall: the line of junction of elytra in Coleoptera.

Suturiform: an articulation soldered together so that only a slight impressed line is visible.

Swarming: the concerted departure from a hive of a large number of worker bees, accompanied by a queen; this forming the nucleus of a new colony.

Swimmerets: gill or plate-like structures in the aquatic larvae of some Neuroptera, serving as oars or organs of locomotion.

Swimming paddles: terminal appendages of mosquito pupae.

Swoked: smoky, suffused with gray or blackish.

Sylvan: species inhabiting forests or woodland areas.

Symbiogenesis: the method of origin of social symbiotic relation among ants and other insects.

Symbiosis: a life relationship existing between different kinds of animals or plants, or between animals and plants: true symbiosis is where both parties to the relation benefit: see also parasitism, commensalism. Among the ants social symbiosis exists in its most highly developed form and distinctive terms have been proposed for the various types of relations:

Calobiosis, is that association in which one species, often only the female, lives in the nest of and at the expense of another species, either for a time, = temporary - or altogether, = permanent calacobiosis.

Cleptobiosis, is where one species of ant lives in or near the nest of another, preying upon its larvae or pupae or stealing the food supply.

Dulosis, is that mingling of colonies which owes its origin to the enslavement of one species by another.

Hamabiosis, is that relation where two species of any insects, one of which may be an ant, live side by side without obvious motive or known advantage to one or both.

Lestobiosis, is where the workers of one ant colony "hold up" those of another species and rob them of the food they are carrying to the nest.

Parabiosis, is where different species of ants form colonies with inosculating galleries, and have their households strangely intermingled, but not blended.

Phylacobiosis, is the relation existing between ants and Termites, the ants living in the doorways of the Termites and functioning as guards.

Synclerobiosis, is an association of two species of ants that usually inhabit independent colonies, for purposes that are not clearly understood.

Trophibiosis, is the relationship between ants on the one hand and aphids, coccids and the like on the other; these species being sought and attended by the ants for their own benefit: see myrmecophily.

Xenobiosis, is where one species of ant lives as a guest in the nest of another, maintaining its own household, and mingling freely with the host species, the two living on terms of mutual toleration.

Symbiotic: species that live together in a state of symbiosis.

Symmetrical: evenly developed on both sides.

Symmetry: that regular arrangement of organs or parts which is capable of division into similar halves or similar radii.

Sympathetic nervous system: applied to the nerves and ganglia of the alimentary canal and sonic other viscera which they innervate; = vagus; visceral nervous system.

Symphily: the relation borne to ants by the true guests which inhabit their nests and are fed and tended: rendering in return some substance or service desired by the ants: see metochy and synechtry.

Symphyla: a group name for apterous species resembling myriapods in appearance, with functional abdominal legs and the genital openings on the last abdominal segment: regarded by some as connecting forms between insects and myriapods, e.g. Scolopendrella.

Symphysis: where two sclerites are joined together by a soft membrane, permitting a slight motion.

Synaptera: originally wingless insects without metamorphosis; the Thysanura.

Synarthrosis: an articulation without motion.

Syncerebrum: the compound brain of insects.

Synchronous: happening at the same time.

Synciput: that portion of the vertex lying between the eyes.

Syncitium: masses of protoplasm with nuclei, found in ovarian tubes; giving rise to ova, nutritive cells or both.

Syndesis: that method of articulation where two parts are connected by a membrane which permits of considerable motion between them.

Synechtry: the relation borne to ants by insects inhabiting their nests in spite of the efforts of the ants to destroy them: see symphily and metochy.

Synista or Synistata: those Neuropterous insects in which the mouth structures are undeveloped, forming an imperfect tubular structure: see elinguata.

Synoecy: the relation that exists between ants and those guests that are indifferent to and tolerated by them:= metochy, and see symphily and synecthry.

Synonym: a name applied to a species or genus that has been previously named and described.

Synonymous: words of different derivation applied to the same conception.

Synthlipsis: the basal constriction of the notocephalon in Notonectids.

Syntype: = co-type; q.v.

Syringe: in Hemiptera, a chamber into which the salivary ducts open and by means of which the secretion is forced forward between the seta or lancets.

System: an order of arrangement.

Systematic: in definite order, or arranged according to a system.

Systole: that regular contraction of the heart that sends the blood outward: see diastole.

T

T. A. line: transverse anterior line; crosses the primaries of certain moths one-third or less from the base: = antemedial line.

Tactile: used for touching; an organ that has the sense of touch.

Taenia: a broad longitudinal stripe.

Taeniate -us: with broad longitudinal markings.

Taenidium -ia: the band or chitinized fibre forming a part of the spiral thread in the trachea of insects.

Tail: an elongated terminal segment of the abdomen: the cauda in plant lice: elongated processes on the secondaries, in some Lepidoptera and Neuroptera.

Tangential: set in or meeting at a tangent; applied to ornamentation and processes.

Tarsal: relating to the tarsi, or feet.

Tarsal lobes: membranous appendages arising from the underside of the tarsal joints in some Coleoptera.

Tarsus -i: the foot; the jointed appendage attached at the apex of tibia. bearing the claws and pulvilli.

Taste cups: specialized pits or cups, with or without a peg or hair, connected with ganglionated nerve cells: occur on the mouth structure and evidence the sense of taste.

Tawny: a brownish yellow, like the color of a tanned hide [pale cadmium yellow + Indian red].

Taxonomical: systematic: relating to classification.

Testate: covered; concealed: also used as = tectiform.

Tectiform: roof-like, sloping from a median ridge, like the primaries of Cicada.

Tegmen: a covering: sometimes used for the anterior wings in Orthoptera and Neuroptera.

Tegmina: the thickened primaries serving as wing covers in Orthoptera.

Tegulae: small, more or less cup-like scales at the base of primaries in many insects; specifically in Hymenoptera: in Lepidoptera, = the patagia or shoulder tippets; but the homology is disputed; also applied to the lappet-like pieces forming the collar: in Diptera, the alulae, q.v.: the latter use is unfortunate and should be abandoned; the first definition should limit the use of the term: see aileron.

Tegument: a covering surface or skin.

Teleodont: applied to those forms of male Lucanids bearing the largest mandibles: see mesodont, amphiodont, priodont.

Telescopic: arranged so that one portion of an organ or process may be drawn into another, like the joints of a telescope.

Telson: a terminal tubercle bearing the anal opening: the anal segment of the insect embryo.

Telum: a spear, or spear-shaped process.

Temple: the posterior part of the gena; behind, before or beneath the eye.

Tempora: the temples.

Temporal margins: in Mallophaga, the lateral margins of the hind head.

Tenaculum: in Collembola, a small organ which holds the furcula in position when at rest: = catch.

Tenant hair: see tenent hair.

Tendo: the anal area of secondaries when it forms a groove for the abdomen: has also been called frenum and frenulum: in Trichoptera, a small elliptical space at base of hind wings near base of anal veins and behind the trochlea.

Tendon: the slender, chitinous plates, bands, strap- or cup-shaped pieces, to which muscles are attached for moving appendages: see apodeme.

Tenent hair: specialized hair adapted for clinging or clasping.

Teneral: that state of the imago just after its exclusion from pupa or nymph, in which neither coloring nor clothing is fully developed.

Tensor: a muscle which stretches a membrane.

Tentacle: a flexible sensory or tactile process; in some cases retractile: usually prefixed by a descriptive term indicating the structure to which it is attached.

Tentacular -um: retractile processes on the larvae of Lepidoptera.

Tentaculate: a margin when fringed with soft tactile processes.

Tentiform: shaped like a tent: see mines.

Tentoria: Diptera; two hollow, cylindrical struts which pass from the ventral border of the occipital foramen to the cheeks.

Tentorium: a chitinous frame-work within the head, upon which the brain rests.

Tenuis: thin, slender; long drawn out.

Terebra: a borer or piercer: an ovipositor fitted for boring or cutting as in sawflies: a mandibular sclerite articulated to the basalis; forms the point of the structure and = the galea of the maxilla.

Terebrant: with an ovipositor fitted for piercing or boring.

Terebrantia: Hymenoptera with sessile abdomen and valved ovipositors: Thripids in which the ovipositor of female is borer-like.

Teres, Terete: cylindric or nearly so.

Tergal: belonging to the primitively upper surface: see dorsal.

Tergal suture: the Y shaped dorsal suture on the head of many insect larvae.

Tergite: the primitively dorsal part of a segment, especially when that part consists of a single sclerite; usually applied to the abdomen.

Tergo-pleural: the upper and lateral portion of a segment.

Tergo-rhabdites: the lower pair of corneous appendages forming the ovipositor in grasshoppers: plates on the inner dorsal surface of the abdominal wall.

Tergum: the primitively upperor dorsal surface whether it consists of one or more than one sclerite and specifically of the abdomen: in Odonata and Orthoptera, applies to thorax as well.

Termen: the outer margin of a wing, between apex and hind or anal angle.

Terminal: situated at the tip or extremity; opposed to basal.

Terminal line: in Lepidoptera, runs along the outer margin of the wings.

Terminal space: the area between the s. t. line and terminal line in certain Lepidoptera.

Terminology: the technical nomenclature of any science.

Termitarium: a nest, natural or artificial, or a colony of Termites.

Terrestrial: living on or in the land; opposed to aquatic.

Tessellated: checkered; tiled more or less like a chess-board.

Test: the secretionary covering of Coccidae, and especially such as are waxy, horny or glassy.

Testaceous: dull yellow brown; tile colored [pale cadmium yellow+burnt sienna].

Testes: the tubular structures in the male, in which the production of spermatogonia, and often also of later stages in the development of the sperm takes place.

Testicular follicles: in the larva, are those structures which in the adult form the tubes composing the testes; in the adult applied also to the tubes forming the testes.

Testudinate -us: resembling the shell of a tortoise.

Tetra-: four: a combining form.

Tetrachaetae: applied to those Diptera in which the mouth structures consist of four longitudinal blades or piercing structures.

Tetradactyle: with four fingers or finger-like processes.

Tetragonal: having four sides or angles: quadrangular.

Tetramera: applied to Coleoptera with four-jointed tarsi.

Tetramerous: having four-jointed tarsi.

Tetrapoda: applied to those butterflies in which the anterior legs are atrophied in whole or in part.

Tetraptera: a term proposed for all insects with four naked, membranous reticulated wings.

Thamnophilous: applied to species living in thickets or dense shrubbery.

Theca: a case or covering: specifically applied to the fleshy covering of the fly-mouth; to the cases of the Trichopterous larvae; to the lower piece of the male genitalia in Homoptera; and to the outer covering of the pupa.

Thelyotoky: parthenogenetic reproduction when the progeny are all females see Arrhenotoly and Deuterotoky.

Thigh: see femur.

Thigmotactic: contact-loving: applied to species that tend to live close together or in touch, one with the other.

Third longitudinal vein: in Diptera (Will.):= radius 5 (Comst.).

Third posterior cell: in Diptera, = 2d medial 2 (Comst.).

Third submarginal cross-nervure: in Hymenoptera (North.):= radius 4 (Comst.).

Thoracic: belonging or attached to the thorax.

Thoracic dorsal bristles: in Diptera, the specialized bristles on the dorsum of the thorax.

Thoracic feet: the jointed legs on the thoracic segments of larvae, as distinguished from abdominal or pro-legs.

Thoracico-abdominal: the first segment of the abdomen when united with the thorax so as to form part of it: =propodeum.

Thoracic pleural bristles: in Diptera, the specialized bristles situated on the pleural region of the thorax.

Thoracotheca: = cytotheca: q.v.

Thorax: the second or intermediate region of the insect body, bearing the true legs and wings: made up of three rings, named in order, pro-, meso-, and meta-thorax: when the pro-thorax is free as in Coleoptera, Orthoptera, and Hemiptera, the term thorax is commonly used in descriptive work for that segment only: in Odonata, where the prothorax is small and not fused with the larger and united meso- and meta-thorax, the term thorax is commonly used for these latter two united, excluding the prothorax.

Thread-plate: an epithelial plate of the embryo from which the terminal threads of the ovarian tubes originate.

Thyridial cell: in Trichoptera: the cell formed by the first fork of median vein; the cell behind Thyridium.

Thyridiate: applied to a wing vein that at one point seems broken so as to permit of a folding or bending; either to pack into a small compass or to enfold the body.

Thyridium -ii: small, whitish or almost transparent spots near the anastomosis of the disc of the wings in some Neuroptera, or in the recurrent veins in the cubital cellule in some Hymenoptera; also the apical margin of the gastrocoeli, often alone visible: in Trichoptera, specifically, a hyaline spot on second fork of median vein.

Thyrsus: a cluster.

Thysanoptera: fringe-winged: an ordinal term, applied to species with four narrow, similar wings, lengthily fringed; mouth parts fitted for puncturing and scraping; metamorphosis incomplete: the Thripids.

Thysanura: fringe-tails; wingless, mandibulate insects without metamorphosis; with anal appendages; body covered with scales; thoracic segments similar.

Tiarate -us: turban or tiara-like.

Tibia -ae: the shank: that part of the leg articulated to the femur basally and which bears the tarsus at the distal end.

Tibial epiphysis: a movable process attached near the base of the inner side of the anterior tibia in many Lepidoptera.

Tibial membrane: in male Cicada, the drum-like vibratory membrane that produces the sound.

Tip: the extremity; the part furthest removed from the base.

Titillator: a small process just below the penis in some Orthoptera.

Tomentose: covered with fine hair, so matted together that particular hairs cannot be separated.

Tomentum: a form of pubescence composed of matted, woolly hair: in Diptera applied to a covering of short, flattened, more or less recumbent, scale-like hair which merges gradually into dust or pollen.

Tongue: an indefinite term, applied usually to the coiled mouth structure of Lepidoptera; the lapping organ of flies; the ligula of bees and wasps and, sometimes also to the hypopharynx of other insects.

Tooth: an acute angulation: a short pointed process from an appendage or margin.

Topomorph -ic: a geographic form, variety or subspecies of a widely distributed species: developed by local environment.

Topotype: is a specimen collected in the exact locality whence the original type was obtained.

Tornal: relating to or concerning the tornus.

Tornus: in Lepidoptera, the junction of the termen and dorsum of wing: = hind or anal angle; q.v.

Torose: swelling into knots or protuberances.

Torpid: lying motionless by reason of cold or other natural conditions that unfavorably affect the organism.

Torqueate: with a ring or collar.

Torquillus: = rotula.

Tortilis: twisted.

Tortulose-us: hump-backed; a surface with a few large elevations: beaded; moniliform.

Tortuose -us: irregularly curved and bent; snake-like.

Tortuous: = Tortuose.

Torulus: the basal socket joint of the antenna upon which the organ is articulated for movement in all directions.

Totidem: in all parts; entirely.

T.P. line: transverse posterior line; crossing the primaries of certain Lepidoptera, two-thirds or more from base: = post medial line.

Trabecula: rounded, lobular masses of the procerebrum, from which arise the stalks bearing the mushroom bodies: a paired movable appendage in front of the antennae in certain bird-lice.

Trachea -ae: the spirally ringed breathing tube or tubes of insects.

Tracheal gills: the flattened or hair-like processes in aquatic larvae through which oxygen is absorbed from the water.

Tracheary: relating to or composed of tracheae.

Tracheate: supplied with trachea: a general term applied to all articulates that breathe by means of spiracular openings into a system of tubular structures that extend to all parts of the body.

Tracheation: the arrangement or system of distribution of trachea.

Tracheoles: the capillary trachea of the adult as they develop in masses in the larva: very small, slender tracheae.

Transection: a cut across, at right angles to the body: transverse section.

Transition zone: is the transcontinental belt in which the austral and boreal elements overlap: it is divided into a humid or Alleghanian area; a western arid area; and a Pacific Coast humid area: all of which see.

Transitory: lasting for a short time only.

Translucent: semi-transparent; admitting the passage of light but not of vision.

Translucid: clear: transparent enough to be seen through.

Transparent: so clear as not to obstruct vision.

Transverse: when the longest diameter is across the body.

Transverse incision: = transverse sulci.

Transverse sulci: the transverse grooves of pronotum in many Orthoptera.

Transverse suture: in Diptera, a transverse groove extending inward from the root of wing and obsolete in the middle of dorsum.

Trapeziform: in the form or shape of a trapezium.

Trapezoid -al: a four-sided plane of which two sides are parallel and two are not.

Tri-: three; a combining form.

Triangle: in Odonata: a small, triangular cell at the junction of cubitus with cubitus 1: a similar cell adjoining it basally is the internal triangle discoidal triangle: cardinal cell; q.v.

Tri-articulate: composed of three joints or articles.

Tribe: a term of classification less than a subfamily: opinionative and ending in ini: but this is not universally adhered to.

Tri-carinate: with three keels or carinae.

Trichogen: a hair-forming hypodermal cell in caterpillars, etc.

Trichoptera: hairy-winged: insects with hairy primaries with many longitudinal veins and cells, covering the broader secondaries which are usually folded lengthwise; mouth mandibulate but rudimentary: head free; thorax agglutinate: metamorphosis complete.

Trichostical bristles: in Diptera, a fan-like row, situated on the meta- pleura: conspicuous in some families.

Trichotomous: divided by threes.

Trichroism: the condition when any given part exhibits three different colors in different individuals of the same species: e.g. in Lepidoptera, the hind wings of certain Heliconids.

Tricuspidate: ending in three points: with three cusps or teeth.

Tridactyle -ous: having three toes or claws.

Trifid: cleft into three parts or ends.

Trigonal: triangular: an area bounded by a triangle.

Trigonate: three-cornered; approximately triangular.

Trigoneutism: where three broods occur in one season.

Trigonulum: in Odonata, = triangle.

Trimera: that series of Coleoptera, in which there are only three tarsal joints present.

Trimerous: species which have the tarsi three-jointed.

Trinomial: that method of nomenclature in which a varietal or subspecific name follows the specific term without an intervening mark or indications of its rank.

Tripectinate: when an antenna has three branches or processes to each joint.

Triquetral: = triquetrous.

Triquetrous: with three flat sides.

Tri-regional: divided into three distinct parts or regions.

Trito-cerebral segment: see second antennal segment.

Trito-cerebrum: the posterior portion of the brain, formed by the ganglion of the third primary segment; also termed labro-frontal lobe.

Tri-undulate: with three waves or undulations.

Triungulin: the first larval stage of a meloid beetle.

Trivial: applied to a name, means specific as opposed to generic, or popular as opposed to technical.

Trivittate: with three stripes or vitta.

Trochalopoda: Heteroptera in which the posterior coxae are nearly globose and the articulation is a ball and socket joint: see pagiopoda.

Trochanter: a sclerite, sometimes divided, between the coxa and femur sometimes fused with the femur.

Trochanterellus: see apophysis.

Trochantine: the basal part of the trochanter when it is two-jointed: in Coleoptera, a piece often present on the outer side of and sometimes movable on the coxa; also the small sclerite connecting the coxa with the sternum in Dytiscidae: in Neuroptera and Trichoptera the posterior separated part of the coxa: in Orthoptera, a narrow longitudinal sclerite between mandible and gena.

Trochiformis: cylindro-conic.

Trochlea: the thickened base of the hind wings in Cicada: in Trichoptera a small elliptical space at base of hind wing behind origin of median vein.

Trochlearis: pulley-shaped; like a cylinder contracted medially.

Trochus: that part of an articulated body inserted between the joints.

Trophi: the mouth parts collectively, including the labrum: see buccal appendages.

Trophobiosis: see Symbiosis.

Tropical: is that faunal region which covers the southern part of the peninsula of Florida, the greater part of Central America, the lowlands of southern Mexico south of the table land, and a narrow strip on each side of Mexico which follows the coast northward into the United States.

Tropico-politan: occurring in all tropical regions.

Trumpets: breathing tubes of mosquito pupae.

Truncate: cut off squarely at tip.

Truncature: the truncation or point squarely cut off.

Truncus: the trunk or thorax.

Trunk: the thorax as a whole: the body.

Tryptic: acting like tripsin, the proteolytic ferment of the pancreatic fluid.

Tube: a slender, hollow, cylindrical body: specifically applied to the anal siphon or respiratory tube of mosquito larvae.

Tubercle: a little solid pimple or small chitinous button: really a ring, which may or may not give rise to a seta.

Tubercles: on the thoracic and abdominal segments of caterpillars are anterior trapezoidal; posterior trapezoidal; lateral; posterior stigmatal; anterior stigmatal; sub-primary subventral; pedal and adventral: all of which see.

Tubercula: an elevated triangular process at the anterior angle of the thorax specifically in Hymenoptera.

Tuberculate -ose: formed like a tubercle: a surface covered with tubercles.

Tubercule -ulum: a small tubercle.

Tuberculiform: shaped like a pimple or tubercle.

Tuberculose -ous: covered or set with tubercles.

Tubulifera: Hymenoptera, in which the terminal segments of abdomen are retracted, but may be extended, tube-like: Thysanoptera in which there is no ovipositor and the terminal segments of abdomen are tubular.

Tubulous -ose: formed like a tube: fistulous.

Tubulus: the slender, flexible abdominal segments forming the ovipositor in Diptera.

Tubus: a term used to designate the corneous base of a ligula: the sheath of the tongue.

Tumescence: a swelling or tumid enlargement: a puffed up area.

Tumescent: a little swollen or puffed up.

Tumid: swollen; enlarged; puffed up.

Tunica intima: the inner layer of the silk glands: an inner lining or membrane.

Tunica propria: a layer of epithelial cells and connective tissue lining the interior of the hind gut: the outer layer of the silk glands: a covering or investing membrane.

Tunicate: composed of concentric layers, enveloping one another: said of antennae when each successive joint is buried in the preceding funnel-shaped one.

Turbinate: top-shaped; nearly conical: differs from pyriform in being shorter and more suddenly attenuated at base: applied to an eye = pillared eye; q.v.

Turgid: swollen.

Turritus: towering: a surface rising cone-like.

Tylo: = tylus; q.v.

Tylus: the anterior central lobe of the head in Hemiptera.

Tympana: the ears in Orthoptera.

Tympanal: applied to organs covered with a tympanum or stretched membrane supposed to function as ears.

Tympanic spiracle: in Diptera, the thoracic spiracle at base of wing.

Tympanules: small openings covered by a membrane, having otoliths and serving as ears.

Tympanum: any membrane stretched like the head of a drum: specifically applied to the membrane covering the auditory organs in Orthoptera.

Type: a unique or single specimen selected from a series and labelled by the describer to represent his name and description: if male or female be added to the label, the specimen typifies that sex, and in case of an erroneous association the male type stands for the species unless the author has specifically designated the other example as representing the name: see also co-type; homotype; meta-type; paratype; topotype.

Typical: the normal or usual form of a species; agreeing with the type form.

U

Uliginous: muddy, or pertaining to mud.

Ulnar: in Homoptera, a wing vein between the radial vein and claval suture; = cubitus: in Orthoptera, = cubitus; q.v.

Ulnar area: in Orthoptera, = median area; q.v.

Ulona: the thick, fleshy mouth parts of Orthoptera.

Ulonata: a Fabrician. term for Orthoptera, based on the character of the mouth structures.

Ultimate: last or final: that larval stage just before pupation.

Ultramarine: an intense deep blue [cobalt blue].

Ultra-nodal sector: in Odonata, runs parallel with and between media 1 and 2, or principal and nodal sectors: = postnodal sector.

Umbilicate: navel-shaped, or resembling a navel.

Umbilicus: a navel, or navel-like depression.

Umbonate: bossed; with an elevated knob in the centre.

Umbone: an embossed, elevated knob situated on humeral angle of elytra.

Umbones: two movable spines on the sides of prothorax in some Coleoptera.

Umbrosa: shaded or clouded: a cloud or shade.

Unarmed: without spurs, spines or armature of any kind.

Unarticulate: not jointed nor segmented.

Unci: thick, hooked processes, forming the borders of the anal opening.

Uncinnate: hooked at the end.

Uncus: in Lepidoptera, Diptera, and elsewhere, the curved book directed downward from a triangular dorsal plate in the male and shielding the penis: the genital hamule.

Undate: wavy or waved.

Undulated: obtusely waved in segments of circles.

Unequal: unlike in size, form, development or other characters.

Ungues: the tarsal claws.

Unguiculate: armed with a hook, nail or claw.

Unguiculus: a small terminal claw or nail-like process.

Unguis: one of the claws at the end of the tarsus: also applied to a short process on the 6th antennal joint in some Aphids.

Ungula: a hoof, claw or talon.

Ungulate: shaped like a hoof.

Uni-: one, a combining form.

Unicolorous: of one color throughout.

Unidentate: with one tooth only.

Uniplicate: with a single fold or line of folding.

Unique: one only: unlike any other.

Unisexual: of one sex only: applied to Aphids and Cynipids where only parthenogenic females are known.

Upper austral zone: is divided into an eastern humid or Carolinian area, and a western arid or upper Sonoran area, which pass insensibly into each other near the 100th meridian: see Carolinian and upper Sonoran.

Upper field: in tegmina, = anal field; q.v.

Upper margin: of tegmina (Thomas), corresponds to the posterior or anal margin of most authors.

Upper median area: see areola.

Upper radial: in Lepidoptera, = media 1 (Comst.), and is vein 5, or the independent, of the numerical series.

Upper sector of triangle: in Odonata, = cubitus 1 (Comst.).

Upper Sonoran faunal area: that arid part of upper austral west of 100th meridian; covers most of plains in eastern Montana and Wyoming, s. w. South Dakota, west. Nebraska, Kansas, Oklahoma and Texas, and east. Colorado and New Mexico; covers plains of Columbia, Malheur and Harney in Oregon and Washington. In California encircles Sacramento and San Joaquin Valleys and forms a narrow belt around Colorado and Mohave deserts. In Utah covers Salt Lake and Sevier deserts. In Idaho the Snake plains. In Nevada and Arizona irregular areas of suitable elevation.

Uranidin: a yellow coloring matter in some Coleoptera and Lepidoptera.

Urceolate: pitcher-shaped; swelling in the middle.

Ureter: the stalk connecting the malpighian tubules, when they form large tufts, with the intestine.

Uric acid: the characteristic nitrogenous excretion of the malpighian or urinary tubules: composition, C5H4N4O3 (von Furth).

Urinary vessels: = malpighian tubules; q.v.: has also been applied by older authors, to anal glands.

Urite: an abdominal segment and, specifically, its ventral portion.

Uromere: any of the abdominal segments of an arthropod.

Uropoda: any of the abdominal feet of arthropods.

Uropygium: the ovipositor when it is a mere extension of the abdominal segments.

Urosome: the abdomen.

Urosternite: the sternal or under piece of the uromeres.

Urticating: nettling; applied to specialized hairs or processes on the bodies of certain caterpillars, which cause a stinging or burning sensation on the skin.

Ustulatus: scorched: applied to a maculation that has the appearance of having been burned in.

Uterus: the vaginal portion of oviduct: the sometimes enlarged portion of the vagina at junction of the oviducts: = calyx, q.v.

Uterus masculinus: a pouch or sac into which the ductus ejaculatorius opens in the Symphyla.

Utriculi breviores: small vesicular sacs connected with the seminal vesicles in crickets and some other insects.

Utriculi majores: large vesicular sacs or tubular structures connected with the seminal vesicles in crickets and some other insects.

Utriculus: a little bag or hollow vesicle.

V

Vacuolate: with vacuoles or small cavities, empty or filled with a watery fluid.

Vagina: the tubular structure formed by the union of the oviducts in the female, opening externally to admit the passage of the egg to the ovipositor: receives the penis of the male in copulation and is sometimes called oviduct: "every part, the office of which is to cover, protect or defend the tongue": "the bivalve coriaceous sheath or cover of the spicula": generally, a sheath.

Vaginata: sheathed: an obsolete ordinal term for Coleoptera.

Vaginate: inclosed in a bivalved sheath.

Vagus: sympathetic nervous system; q.v.

Valgate: enlarged at bottom: club-footed.

Valve or Valvulae: the expanded plate-like galea of the maxilla in many Hymenoptera.

Valve: a small, transverse or triangular piece behind the last full ventral segment, at base of plates in male Jassidae and allies.

Valves: in Orthoptera, the corneous pieces of the ovipositor:= corniculi in Lepidoptera, sometimes used to = harpes; q.v.

Valvula = vagina in its application to Dipterous mouth parts.

Valvulae: in Hymenoptera, branches of the genital forceps of male.

Valvular: when two parts join so as to form a valve between them.

Valvular process: in Odonata, a slender, unjointed process at the apex of each genital valve.

Valvule: any small, valve-like process.

Variation: a departure in color or form, from the normal: the sum of the departures from a mean type of any species: it is continuous when there is no break between the extremes; discontinuous when there are gaps without intermediate forms.

Variety: any departure from the normal type of a species which, while retaining the specific characters, is yet recognizably different because of climatic, seasonal or other influences; may occur with the type form or as a geographical race.

Variola: a deep, rounded impression with defined edges.

Variolate -ose: with large, rounded impressions like pock-marks.

Vas deferens: = vasa deferentia, q.v.

Vasa deferentia: tubes from the seminal vesicles or testes of each side, which usually unite into a single ductus ejaculatorius; q.v.

Vasa varicosa: the malpighian tubules.

Vascular: relating to the blood-vessels or circulatory system.

Vasiform orifice: in Aleurodidae, an ovate, triangular or semicircular opening on the dorsum of the last abdominal segment.

Veinlets: in Orthoptera, are the minute transverse ribs or ridges between the longitudinal veins.

Veins: the chitinous, rod-like structures supporting the wings, and especially those extending longitudinally from base to the outer margin nerves nervures; nervules.

Velum: a membranous appendage of the spurs at the apex of anterior tibia in bees a broad process at inner end of fore tibia.

Velum penis: the thin membranous covering of the male intromittent organ also applied to ether covering or shield-like structures of the penis.

Velutinous: velvety: clothed with dense, soft, short hair, like velvet.

Vena: a vein.

Vena dividens: that longitudinal vein of secondaries that marks the beginning of the anal area: = anal 1 (Comst.).

Vena plicata: on the wings of Dermaptera, the vein around which the folding occurs.

Vena spuria: = spurious vein: q.v.

Venation: the system of chitinous frame-work supporting the wings: in Lepidoptera, the veins are usually referred to by numbers which are as follows: on primaries: 1 = anal; 2 = cubitus 2; 3 = cubitus 1: 4 = media 3; 5 = media 2; 6 = media 1; 7 = radius 5; 8 = radius 4; 9 = radius 3; 10 = radius 2; 11 = radius 1; 12 = subcosta: on secondaries: 1, 1a, 1b = anal; 2 = Cubitus; 3 = cubitus 1; 4 = media 3: 5 = media 2; 6 = media 1; 7 = radius 1; 8 = subcosta. See plate III for typical venations of all orders.

Venter: the belly: under surface of abdomen as a whole and of each ring.

Ventose: inflated; puffed out.

Ventrad: extending or directed toward the under side.

Ventral: pertaining to the under surface of abdomen: in Diptera, that face of the leg which is inferior when laterally extended.

Ventral chain: refers to the series of ganglia of the nervous system.

Ventral comb: in Trichoptera, a transverse row of fine teeth on venter.

Ventral diaphragm: is a fine membrane covering the central nerve cords and ganglia: also called ventral heart.

Ventral heart: = ventral diaphragm, q.v.

Ventral plate: a thickening of the blastoderm of an egg from which the embryo, but not the amnion or serosa is formed.

Ventral scale: in Diaspinae, the under part of the puparium, interposed between the insect and the plant.

Ventral tube: in Collembola, a tube or tubercle proceeding from the ventral side of the first abdominal segment.

Ventricose: with a big belly: distended; inflated.

Ventriculus: the true stomach, = chylific ventricle; q.v.

Ventri-meson: the middle line of the ventral surface of the body.

Ventro-cephalad: toward the lower side and anteriorly.

Ventro-dorsad: extending from belly to back.

Venules: the branches of the main veins.

Vermian: worm-like.

Vermicular: worm-like, tortuous: resembling the tracks of a worm.

Vermiculate: worm-like in form: a marking with wormlike tracings.

Vermiform: worm-shaped.

Vernal: appearing in spring.

Vernantia: the molting or shedding of the skin.

Verriculate: with thick-set tufts of parallel hairs.

Verricule: a dense tuft of upright hairs.

Verrucose: having little hard lumps or wart-like elevations.

Versatile: moving freely in every direction.

Versicolored: with several colors, indeterminately restricted.

Vertex: the top of the head between the eyes, front and occiput: in bees, that part of the head adjacent to and occupied by the ocelli: in Notonectids, "the imaginary anterior margin of the notocephalon."

Vertexal: occurring on or near the vertex, or directed toward it.

Vertical cephalic bristles: in Diptera, are two pairs, inner and outer, inserted more or less behind the upper and inner corner of the eye; erect, or the inner pair convergent, the outer pair divergent.

Vertical margin: in Diptera, the limit between front and occiput.

Vertical triangle: in male Diptera, the small triangle upon which the ocelli are situated; limited behind by vertex, in front by eyes.

Verticil: one of the whorls of long fine sensitive hair arranged symmetrically on the joints of the antennae in certain Diptera.

Verticillate: placed in whorls: antennae in which the joints have a circle of long, fine hair as in Cecidomyiids.

Vesicant: blistering: able to produce a blister.

Vesicle of penis: in Odonata, a sac with chitinous walls, attached to the sternum behind the penis.

Vesicles: little sacs, bladders or cysts: applied to extensible organs producing odors or secretions, as in some beetles and caterpillars.

Vesicular: bladder-like; beset with spherical prominences.

Vesicula seminalis: see seminal vesicles.

Vestibule: the space around the ovipositor formed by the projecting margins of the surrounding segments: the space between the occluding structure of the spiracle and the valve opening into the trachea itself.

Vestigial: small or degenerate: only a trace or remnant of a previously functional organ.

Vestiture: the surface clothing, whether of a hairy or scaly character.

Vexhillum: in Hymenoptera, an expansion on the tip of tarsi of certain fossorial groups.

Vibrant: having a rapid motion to and fro.

Vibratile: formed for vibratory motion: used to express the almost continual movement of the antennae of some Hymenoptera, and the wings of some Diptera.

Vibrissae: curved bristles or hairs in some Diptera, situated between the mystax and the antenna: whiskers.

Villi: soft hairs or papillate processes: plural of villus, q.v.

Villose -ous: soft-haired or clothed with soft, short hair.

Villus: a short, hair-like or papillate process on the surface of certain absorbent and sensory organs.

Vinous: wine-color: a deep, transparent red-brown. like claret [purple madder].

Violaceous: violet colored: a mixture of blue and red [violet carmine].

Virescent or Viridescent: greenish or becoming green.

Viridis: green, like verdigris [French blue + chrome yellow + white].

Viscera: the internal organs of the body.

Visceral: relating or attached to the viscera.

Viscid: sticky: covered with a shiny, resinous or greasy matter.

Viscous: thick, sticky or semi-fluid.

Vis formatrix: the creative or formative force.

Vitelligenous: producing the vitellus or yolk: said of certain cells in the ovaries, believed to have that function.

Vitelline -us: yellow, with a slight tinge of red, like yolk of an egg.

Vitelline membrane: the delicate tissue surrounding the yolk of an egg.

Vitreous: glassy; transparent.

Vitta: a longitudinal, colored line.

Vitta frontalis: = frontal stripe: q.v.

Vittate: striped.

Viviparous: applied to insects which bear living young.

Vocal cords: specialized organs on the thoracic spiracles of Diptera, by means of which they produce a humming or singing sound.

Volant: flying or capable of flight.

Vulgar: common; not conspicuous: obscure in appearance and abundant in number.

Vultus: face: that part of head below front and between the eyes.

Vulva: the orifice of the vagina in the female.

Vulvar lamina: in Odonata, the posterior margin of sternum of segment 8.

Vulvar scale: = v. lamina.

W

Wart: a spongy excrescence, more or less cylindric, with a nearly truncated tip: the enlarged, common base of a group of seta: in Trichoptera, a pitted elevation.

Wax: a ductile substance excreted by bees and other insects from glandular structures in various parts of the body, used in building cells or in forming a protective covering.

Wax-cutter: the pincer-like structure formed by the hind tibia and metatarsus in social bees.

Wax-glands: any glands in any part of the body which secrete a waxy product in either a scale, string or powder: in Coccidae, the circumgenital and parastigmatic glands; q.v.

Wax-pincer: = wax cutter.

Wax-scale: one of the scales secreted in the wax pocket or gland of a worker bee.

Whitlows: = paronychia; q.v.

Whorl: a ring of long hair arranged around a centre, like the spokes around the hub of a wheel.

Wing, Wings: membranous reticulated organs of flight; one pair, the primaries, attached to the meso-thorax; the other, the secondaries, attached to the meta-thorax.

Wing covers: those parts of the chitinous cuticle of larvae, nymphs or pupae which cover the rudiments of the wings of the imago: the forewings of an imago when they are thicker than the hind wings and cover them when at rest: see elytra; tegmina.

Wings of the heart: the series of diagonal and other muscular fibres above the diaphragm in the pericardial cavity: see pericardial diaphragm.

Wing cells: areas inclosed by veins: reference should be had to the figures illustrating venation and to the special terms applied to the cells.

Winglets: small, concavo-convex scales, generally fringed at tip, under the base of the elytra in Dytiscidae.

Wing-pads: undeveloped wings of pupa or nymph.

Wing-scale: in Hymenoptera, = tegula; q.v.

Workers: the undeveloped females in the social Hymenoptera; also those sexually undeveloped Termites that are not soldiers.

X

Xanthophyll: the yellow of autumn leaves; one of the substances found in the blood of insects.

Xenobiosis: see symbiosis.

Xerophilous: applied to species living in dry places.

Xylophaga: wood-eaters: applied in several orders.

Xylophagous: feeding in or upon woody tissue.

Xyphus: a spinous or triangular process of the meso-sternum in many Hemiptera, and some other insects.

Y

Yellow: used without modification is sulphur or lemon yellow.

Yolk: the nutritive matter of an egg as distinguished from the living, formative material; = deutoplasm.

Z

Zona: a belt or zone; as of distribution.

Zonite: = arthromere or somite; q.v.

Zoönite or Zoönule: = zonite.

Zygoptera: those Odonata, having the fore and hind wings subequal in width, venation comprising a quadrilateral, not a triangle; nymphs with caudal tracheal gills.

BIBLIOGRAPHY

Adams, Jean, ed. (1992). *Insect Potpourri: Adventures in Entomology.* St. Lucie Press, Delray Beach, Fla.

Akre, R. D. G. S. Paulson, E. P. Catts. (1992). *Insects Did It First.* Ye Galleon Press: Fairfield Washington.

Bates, Marston. (1949). *The Natural History of Mosquitoes.* Macmillan, New York.

Comstock, John Henry. (1903). *Insect Life.* D. Appleton and Company, New York.

Coulson, R.N. and J.A. Witter (1984). "Arthropod pests in recreational areas." In: *Forest Entomology.* John Wiley & Sons, New York.

Dethier, Vincent G. (1992). *Crickets and Katydids, Concerts and Solos.* Harvard Univ. Press, Cambridge, Mass.

———. (1962). *To Know a Fly.* Holden-Day, San Francisco.

Ekbom, K.A. (1938). Der praesenile Dermatozo-enwahn. *Acta Psychiatr. Neurol. Scand.* 13: 227–259.

Evans, Howard Ensign. (1985). *The Pleasures of Entomology.* Smithsonian Institution Press, Washington, D.C.

Gordon, David George. (1996). *The Compleat Cockroach.* Ten Speed Press, New York.

Hogue, C.L. (1987). "Cultural entomology." *Ann. Rev. Entomol.* 32: 181–99.

Matthews, E.G. (1975). "The Mediterranean beetle Blaps polychresta Forskal in South Australia." *The South Australian Naturalist.* 49: (3) 35-39.

Moffet, B.S. (1979). "Even creepy bugs useful, expert says." *Los Angeles Times,* Thurs, 30, Part 1-C 7.

Westwood, J.O. (1839). *An introduction to the modern classification of insects.* Longman, Orme, Brown, Green and Longmans, London.

INDEX